W9-CLF-318

ALVAREZ

BOOKS IN THE ALFRED P. SLOAN FOUNDATION SERIES

Disturbing the Universe *by Freeman Dyson*
Advice to a Young Scientist *by Peter Medawar*
The Youngest Science *by Lewis Thomas*
Haphazard Reality *by Hendrik B. G. Casimir*
In Search of Mind *by Jerome Bruner*
A Slot Machine, a Broken Test Tube *by S. E. Luria*
Rabi: Scientist and Citizen *by John Rigden*
Alvarez: Adventures of a Physicist *by Luis W. Alvarez*

THIS BOOK IS PUBLISHED AS PART
OF AN ALFRED P. SLOAN FOUNDATION PROGRAM

ALVAREZ

Adventures of a Physicist

LUIS W. ALVAREZ

Basic Books, Inc., Publishers

NEW YORK

Library of Congress Cataloging-in-Publication Data

Alvarez, Luis W., 1911–
Alvarez: adventures of a physicist.

(The Alfred P. Sloan Foundation series)
"Published a part of an Alfred P. Sloan Foundation program"—P. 2.
Includes index.
1. Alvarez, Luis W., 1911– . 2. Physicists—United States–Biography. I. Alfred P. Sloan Foundation. II. Title. III. Series.
QC774.A49A3 1987 530'.092'4 [B] 86–71499
ISBN 0–465–00115–7

Printed in the United States of America
Designed by Vincent Torre
87 88 89 90 HC 10 9 8 7 6 5 4 3 2

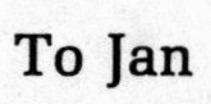

To Jan

CONTENTS

PREFACE TO THE SERIES

THE Alfred P. Sloan Foundation has for many years had an interest in encouraging public understanding of science. Science in this century has become a complex endeavor. Scientific statements may reflect as many as four centuries of experimentation and theory, and are likely to be expressed in the language of advanced mathematics or in highly technical terms. As scientific knowledge expands, the goal of general public understanding of science becomes increasingly difficult to reach.

Yet an understanding of the scientific enterprise, as distinct from data, concepts, and theories, is certainly within the grasp of us all. It is an enterprise conducted by men and women who are stimulated by hopes and purposes that are universal, rewarded by occasional successes, and distressed by setbacks. Science is an enterprise with its own rules and customs, but an understanding of that enterprise is accessible, for it is quintessentially human. And an understanding of the enterprise inevitably brings with it insights into the nature of its products.

Accordingly, the Sloan Foundation has encouraged some outstanding and articulate scientists to set down personal accounts of their lives in science. The form these accounts take has been left to each author: an autobiographical approach, a series of essays, a description of a particular scientific community. The word *science* has not been construed narrowly, but includes technology, engineering, and economics as well as physics, chemistry, biology, and mathematics.

The Sloan Foundation expresses its appreciation of the great contribution made to the program by its advisory committee. The committee has been chaired since the program's inception by Robert Sinsheimer, Chancellor of the University of California, Santa Cruz.

Present members of the committee are Simon Michael Bessie, Co-Publisher, Cornelia and Michael Bessie Books; Howard Hiatt, Professor, School of Medicine, Harvard University; Eric R. Kandel, University Professor, Columbia University College of Physicians and Surgeons and Senior Investigator, Howard Hughes Medical Institute; Daniel Kevles, Professor of History, California Institute of Technology; Robert Merton, University Professor Emeritus, Columbia University; Paul Samuelson, Institute Professor of Economics, Massachusetts Institute of Technology; and Stephen White, former Vice President of the Alfred P. Sloan Foundation. Previous members of the committee were Daniel McFadden, Professor of Economics, and Professor Philip Morrison, Professor of Physics, both of the Massachusetts Institute of Technology; Mark Kac (deceased), formerly Professor of Mathematics, University of Southern California; and Frederick E. Terman, Provost Emeritus, Stanford University. The Sloan Foundation has been represented by Arthur L. Singer, Jr., Eric Wanner, and Sandra Panem. The first publisher of the program, Harper & Row, has been represented by Edward L. Burlingame and Sallie Coolidge. This volume is the second edition to be published by Basic Books, represented by Martin Kessler and Richard Liebmann-Smith.

—Albert Rees
President
Alfred P. Sloan Foundation

INTRODUCTION

ALMOST EVERYONE who writes an autobiography feels a need to explain why he indulged in such egocentric behavior. I realized in 1969 that my two youngest children, then three and five years old, would know little of me if I were to die suddenly, at my then age of fifty seven, as my friend and mentor Ernest Lawrence had. So for them I began dictating my life story, with the strong conviction that I would never take the time to prepare it for publication.

The work went slowly until I started on the events that turned me into a physicist. I had never seen these matters described in such detail in any physicist's autobiography. I shared them with my closest young colleague, Rich Muller. Rich found them to be a handbook of sorts on how to become an experimental physicist. When my son Walt joined the Berkeley geology faculty, I made copies of my transcribed dictation for both him and Rich. For many years only three people knew that I was writing my memoirs—my wife, Jan, was the third.

Several years ago my former student Peter Trower proposed putting together a festschrift for me, reproducing my most important papers, with comments by former colleagues. When I agreed, Peter and I spent many hours talking about the work and events of my long career. In effect, Peter had joined the family, so I shared my rough manuscript with him. He thought it would interest more people than I had assumed, and proposed that it be published. Because I didn't feel I had it in me to cut out the rambling asides that my dictating had produced, I asked Peter to be my editor.

He did an enormous amount of work turning my loquacious drafts into this book. I wrote every substantive word, but Peter has excised and has provided connective tissue and organization. Even though some of my most treasured prose has fallen on the cutting-room floor, I think he's shown excellent judgment. Peter also made the necessary arrangements with the Alfred P. Sloan Foundation and with our publisher. So this book is a joint effort and wouldn't exist if each of us hadn't done the things the other couldn't. After Peter and I had done our best, the manuscript was still too long. The Sloan Foundation supported our efforts then by obtaining the services of Richard Rhodes, a professional writer with extensive knowledge of the atomic-bomb project. Dick has managed to cut the manuscript to a length that is acceptable to the publisher. I have thoroughly enjoyed working with him, as I had with Peter, and if this book gives any pleasure to its readers it will be because all three of us gave it our best shots.

Parts of the book may be too technical for some readers; such parts can easily be skipped. We want young scientists to understand how scientists think and act, and we want to supply historians of science with more source material. Some sections of Bram Pais's wonderful biography of Einstein, *'Subtle is the Lord'* (Oxford University Press, 1982), were mathematically way over my head, but I could skim them without losing the story line. I hope my readers will be able to skip any technical details here in the same way and share the human excitement of the research laboratory with me and my colleagues.

ALVAREZ

PROLOGUE

The Hiroshima Mission

THE ATOMIC BOMBS that ended World War II, Little Boy and Fat Man, were delivered to Hiroshima and Nagasaki from an airfield on Tinian, a small island in the Mariana chain, between Guam and Saipan. An American physicist, thirty-four years old in 1945, I was there at the time and flew the first of the two historic missions, the leader of a small group responsible for monitoring the energy of the explosion.

Tinian resembles Manhattan in size, shape, and orientation. The Seabees who constructed its airfields noticed the resemblance and made it complete with signing; Tinian's southern tip became "The Battery," the road up the west side "Riverside Drive," the road from the Battery to North Field "Broadway," the headquarters at Forty-second and Broadway "Times Square." North Field was then the largest airport in the world. It covered four square miles, with wide east-west runways two miles long and hundreds of parking revetments for B–29's. In the last year of the war, flights of up to five hundred bombers departed two or three times a week from North Field, a second field to the southwest, another on Saipan, and a fourth on Guam to drop thousands of tons of incendiary and high-explosive bombs on Japan. The death and destruction visited by these conventional bombing missions far exceeded the toll taken by the Hiroshima and Nagasaki bombs; by the end of the war Twentieth Air Force Commander Curtis LeMay's crews had burned out fifty-eight Japanese cities and accounted for hundreds of thousands of lives. The atomic bombs were spectacular, which is why they ended the war, but they were not more destructively effective than conventional incendiary bombing as practiced in Japan.

Robert Oppenheimer, the director of the secret wartime laboratory at Los Alamos, New Mexico, where the Little Boy and Fat Man atomic bombs were designed, sent me and several colleagues to Tinian late in July 1945 to monitor the parachute-deployable pressure gauges we had designed to measure the energy of the two explosions. I had gone to Los Alamos after working on the development of radar in the United States and in Great Britain, had invented a new kind of detonator for the Fat Man plutonium bomb, and had then sought assignment that would carry me closer to the front lines in the Pacific war. The uranium 235 used in Little Boy had been separated so laboriously from the more common uranium isotope 238 and at such enormous expense that a bomb of that type had not been tested. We were confident it would work, and Robert wanted to know its explosive yield. Hence my assignment.

When we arrived at Tinian, Larry Langer, one of my Los Alamos poker partners, met our Green Hornet transport plane and delivered us to the tents that would be our bedrooms for the next two months. We were billeted in the 509th Group headquarters area, just south of North Field, a miniature city complete with sleeping areas, bathing facilities, a large mess hall, an outdoor movie theater, and an officers' club. The theoretical physicist Bob Serber had managed to join our Pacific venture. He and I shared a tent; my young colleagues Larry Johnston and Harold Agnew shared another one directly across the street.

Langer gave us a quick jeep tour and delivered us to the technical area, closely guarded Quonset huts at the northern edge of the airfield. Los Alamos had supplied our large Quonset hut with a primitive air conditioner, really a dehumidifier, which kept us comfortable even though the outside temperature was often over eighty degrees Fahrenheit and preserved our equipment from deteriorating in the salty, humid air. Driving daily between our quarters and the tech area, we passed a military cemetery with hundreds of crosses and Stars of David marking the graves of the Americans who died taking

Tinian from the Japanese. Sobering as those graves were, I was heartsick to think how many more would be dug when the invasion of the Japanese home islands began in the fall. Enormous quantities of materiel were then being amassed in warehouses, from Okinawa to Tinian, in preparation for that invasion. Among that materiel, I was told, were shiploads of caskets waiting to be filled with the bodies of American servicemen. Such stark realities drove us in our work.

Our Quonset hut was fitted with workbenches and test equipment for calibrating our microphones. The plutonium core of Fat Man, the Nagasaki bomb, occupied a shelf a few feet from where I worked once it arrived. A small sphere, plated with one-thousandth of an inch of nickel to absorb alpha particles, it was stored in an aluminum box with cooling fins to dissipate the heat generated by its alpha decay. Each of us on several occasions held this ball in our bare hands, feeling its warmth. (I continue to read nonsense about the dangers of handling plutonium. It can certainly be handled without the frequently mentioned lead-lined gloves.) An Army guard stationed beside the core protected it with his life. He obviously didn't know what he was guarding. It was good duty, sitting all day in an air-conditioned Quonset hut. When Hiroshima was announced, we told him he was guarding material of the kind that had just completely destroyed a city. After that he guarded the aluminum box with great respect.

The scheduling of the first atomic bombing was determined primarily by the availability of the necessary U-235 for Little Boy, a mechanism that worked something like a cannon, firing a "bullet" of uranium up a barrel into uranium target rings to assemble a critical mass. The question of how to transport that extremely scarce nuclear material to Tinian had occasioned serious debate. The U-235 bullet came out by sea, on the cruiser *Indianapolis,* along with the nonnuclear components of the Little Boy bomb; the three target rings were flown out on Green Hornets. Sea delivery was assumed to be more reliable (the target rings were ready only on July 26, the day the *Indianapolis* arrived at Tinian), but that assumption almost proved to be disastrously wrong. Only three days after its arrival, when it was steaming toward Okinawa to join the invasion task force assembling there, the cruiser was sunk by a Japanese submarine working in the Philippine Sea. If the *Indianapolis* had been sunk with Little Boy aboard, the war could have been seriously prolonged.

Time passed quickly for us. Little Boy was ready on July 31 and would have flown August 1 if a typhoon approaching Japan hadn't delayed it. Nightly briefings began about August 3. For two nights the weather scrubbed the mission, and we all went to bed. By August 5 the typhoon had passed. Little Boy was loaded that day with much ceremony. Paul Tibbets, the 509th commander who would fly the mission, had newly christened the delivery B–29 *Enola Gay,* his mother's Christian names. The plane taxied over a loading pit, where a hydraulic lift supported the bomb on its detachable cradle. Everyone signed a rude message to the emperor on the bomb case. The lift elevated the bomb into the bomb bay. After that ceremony we went to the movies, then filed into the briefing room and for the first time learned the name of our target—Hiroshima.

In July 1943 I had attended the briefing for the largest conventional bombing raid in history, the bombing of Hamburg by the British Royal Air Force. Now I was being briefed for an equivalent raid by a single plane. Then I had been a spectator; now I was a participant. The difference occasioned very different feelings.

In our crew lockers we collected our personal equipment: a parachute, a one-man life raft, a flak helmet, a flak suit, and emergency rations. We bundled this gear into waiting trucks and rode out to the 509th revetments. Only two planes had initially been scheduled for this mission. After the test of the plutonium bomb at Trinity Site, in the New Mexican desert, where the method worked well, Robert Oppenheimer had decided we could learn something about the bomb's energy release by photographing the expanding fireball with a Fastax camera. The 509th then laid on a third plane, and Bernie Waldman surprised us by showing up at the last minute with two Fastax cameras and thousands of feet of film.

I flew in the *Great Artiste,* piloted by Chuck Sweeney. On so historic an occasion I should probably have been wide awake, but it was well after midnight and I was tired. The forward B–29 compartments connected to the rear, where our equipment was installed, via a tunnel. I found some cushions, crawled up into the tunnel, and fell sound asleep before the *Great Artiste* started its takeoff roll.

Larry Johnston woke me several hours later to ask me what I considered an unimportant question. I wasn't happy to be disturbed, but I noticed that the sun was up and realized I had work to do. We had been flying at low altitude to avoid the head winds of the jet stream and to save fuel. Before we began our climb and the plane was pressurized, we had to deploy the receiving antenna for our parachute gauges through the rear compartment floor.

That done, we watched our progress up the island chain on the radar repeater screen. By the time we saw the coast of Shikoku a hundred miles ahead, we had checked our radio transmitters and were seeing signals on our oscilloscopes. Approaching the coast, we donned flak suits and arranged flak mattresses to sit on. I decided not to wear a parachute; if we were shot down, I didn't want to be captured.

We were too busy to look out the windows as we crossed Shikoku, but we watched our progress on radar across the Inland Sea and over Honshu. Then we were too busy even to watch the radar. Sweeney's comments on the intercom alerted us that we were closing with the target. The tone signal started from the *Enola Gay.*

Abruptly it stopped. The bomb took forty-five seconds to drop thirty thousand feet to its detonation point, our three parachute gauges drifting down above. For half that time we were diving away in a two-g turn. Before we leveled off and flew directly away, we saw the calibration pulses that indicated our equipment was working well. Suddenly a bright flash lit the compartment, the light from the explosion reflecting off the clouds in front of us and back through the tunnel. The pressure pulse registered its N-shaped wave on our screens, and then a second wave recorded the reflection of the pulse from the ground. A few moments later two sharp shocks slammed the plane.

After we secured our equipment, we left our cramped quarters and looked out the window for the first time over Japan. By then Sweeney was heading back toward Hiroshima, and the top of the mushroom cloud had reached our altitude. I looked in vain for the city that had been our target. The cloud seemed to be rising out of a wooded area devoid of population. My friend and teacher Ernest Lawrence had expended great energy and hundreds of millions of dollars building the machines that separated the U-235 for the Little Boy bomb. I thought the bombardier had missed the city by miles—had dumped Ernest's precious bomb out in the empty countryside—and I wondered how we would ever explain such a failure to him. Sweeney shortly dispelled my doubts. The aiming had been excellent, he reported; Hiroshima was destroyed.

We flew once around the mushroom cloud and then headed for Tinian. Japan looked peaceful from seven miles up. After we crossed the coastline, we stripped off our flak gear and went forward to talk to the crew. We could have beaten the *Enola Gay* to Tinian since we had carried no payload and had more fuel on hand, but Sweeney throttled back to let his commander go in first. We followed Tibbets an hour behind; I used the time to write a long letter to my four-year-old son, Walt, for him to read when he was older. I was thinking about the consequences of the bombing I had just witnessed, as these two paragraphs attest:

> The story of our mission will probably be well known to everyone by the time you read this, but at the moment only the crews of our three B–29's, and the unfortunate residents of the Hiroshima district in Japan are aware of what has happened to aerial warfare. Last week the 20th Air Force, stationed in the Marianas Islands, put over the biggest bombing raid in history, with 6,000 tons of bombs (about 3,000 tons of high explosive). Today, the lead plane of our little formation dropped a single bomb which probably exploded with the force of 15,000 tons of high explosive. That means that the days of large bombing raids, with several hundred planes, are finished. A single plane disguised as a friendly transport can now wipe out a city. That means to me that nations will have to get along together in a friendly fashion, or suffer the consequences of sudden sneak attacks which can cripple them overnight.
>
> What regrets I have about being a party to killing and maiming thousands of Japanese civilians this morning are tempered with the hope that this terrible weapon we have created may bring the countries of the world together and prevent further wars. Alfred Nobel thought that his invention of high explosives would have that effect, by making wars too terrible, but unfortunately it had just the opposite reaction. Our new destructive force is so many thousands of times worse that it may realize Nobel's dream.

Flying the Hiroshima mission was certainly one of the most somber and impressive experiences of my life, but it was not my first adventure, nor would it be my last.

ONE

Beginnings

PEOPLE have sometimes been surprised that a tall, ruddy blond should bear the name Luis Alvarez. Although my paternal grandfather was born in northern Spain, my mother's father came from Ireland. All four of my grandparents were adventurous and raised their families thousands of miles from home.

Grandfather Alvarez, who had run off to Cuba in his teens, arrived in California in the 1870s and within a decade gained some measure of prosperity. He acquired a block of downtown Los Angeles. When he realized that there was not enough water in the area to support a community much larger than the village of Los Angeles, he became discouraged, sold his real estate, and enrolled in San Francisco's Cooper Medical College, the forerunner of the Stanford Medical School. He married while a student, and my father, Walter Alvarez, was born. Soon after his graduation the family moved to the kingdom of Hawaii, where my grandfather worked as a government physician responsible for the remote northwest shore of Oahu. My father grew up in the village of Waialua, in isolation from books and libraries. When the family moved to Honolulu, in 1895, he reacted like a starving man at a banquet; the public library became a second home.

My mother's parents founded a missionary school in Foochow,

China. Having grown up at the mission, she was ignorant of the ways of American children until she was sent back to attend high school in Berkeley, California. When I was a small boy, my mother told me stories of her Chinese schoolmates and the courtesy aunts and uncles who lived in the mission compound. But I remember little of my maternal grandmother and nothing of my grandfather.

My father enrolled in Cooper Medical College immediately upon graduation from high school. He interned at San Francisco General Hospital, experiencing during that training the devastating earthquake that occurred in 1906.

My mother, Harriet Smyth, graduated from the University of California that year; the next year she and my father were married. He took up lucrative employment as physician to a mining company, and they moved immediately to Cananea, a small Mexican mining town twenty-five miles below the Arizona border. I once thought that my father must have been a very poor student to have had to resort to employment as a company physician. Later I learned that he had graduated first in his class; he went to Mexico because the job was extremely well paid. Three years in the Wild West produced my sister Gladys. Then the family returned to San Francisco. Dad went to Harvard University for a time to train as a research physiologist in the laboratory of Walter Cannon. I was born in San Francisco in 1911. My younger siblings, Bob and Bernice, followed in 1912 and 1913.

We saw very little of my father in those days. He was busy with two strenuous jobs. His mornings were devoted to research; in the afternoon he worked in private practice to support his family. We children were usually in bed when he came home from his office. But Sunday afternoons, after mother had ferried us all to church and back by streetcar, we spent with Dad. Those family drives in the country were among his few recreations.

I attended the San Francisco Pan-American Exposition with my father in 1915. He remembered later that I was fascinated by the Machinery Hall exhibits, the beginning of my lifelong interest in hardware. Around that time I was confined to bed for a year because it was suspected that I had heart trouble. My mother, who had trained as a grammar school teacher, taught me through the second grade. I skipped the first half of the third grade and so was always one year younger than my classmates, which allowed me to start my life as a physicist a year earlier than most of my contemporaries.

My early exposure to science was substantial. I often accompanied my father on Saturdays to the two rooms at the Hooper Foundation where he conducted his physiological research. His specialty was the activity of the stomach and the intestines, a subject on which he later wrote a definitive work. In his laboratory I usually encountered an anesthetized dog lying on its back in saline solution, its abdomen opened to expose its intestines, threads connecting the viscera to styluses that traced lines on a revolving smoked drum. The kymograph traces recorded the excitation waves of intestinal activity as they progressed down the gut. I didn't find this study at all interesting, but I was fascinated by the electrical equipment in the adjoining room. By the time I was ten, I could use all the small tools in my father's little shop, measure resistances on a Wheatstone bridge, and construct circuits.

In my eleventh year my father gave me a *Literary Digest* article that described how to make a crystal radio. I had never heard a radio; they were used up to that time primarily for rescues at sea. The *Literary Digest* design depended on a cylindrical ice cream carton, around which I wound one hundred turns of shellacked copper wire that served as an inductor. A crystal of galena with a cat's-whisker point contact—a solid-state detector—rectified the radio frequency signals from the inductor and permitted the signal to vibrate the diaphragms of a pair of earphones. (The vacuum tube eventually replaced the galena crystal in radio technology, but the crystal reappeared during the Second World War as the microwave mixer. Attempts to improve the performance of these solid-state rectifiers led directly to the transistor.) My father and I finished our receiver on a Sunday evening. I took a sick day Monday listening to the Bay Area radio stations, which were just beginning to broadcast music. For a while after that everyone who ventured into our house was bustled upstairs to marvel at the wonders of my radio set.

My father recuperated from his heavy responsibilities by going on the annual Sierra Club High Trip. In my twelfth summer I went with him. For three wonderful weeks we hiked fifteen or twenty miles a day on the Muir Trail of the Sierra Nevada, almost as deserted then as it is crowded now. At one point my father, three other men, and I were camped in the Evolution Basin fifty miles from the nearest town when I complained of a sore stomach. My father pulled up my shirt, poked at my abdomen, and instructed me to lie down by the campfire. Only an hour later, when I made a miraculous recovery, did

I learn that he had been boiling knives, forks, and spoons to prepare to remove my appendix. He hadn't operated on a human patient since his days in Mexico, but he was so confident of his diagnosis that he was ready to open me up and cut out the inflamed organ in the middle of the Sierra wilderness.

On a later occasion his medical intervention was more crucial. I spent two weeks for several summers at a Boy Scout camp in the redwoods north of San Francisco. One stay at summer camp was cut short by a family wedding. At camp I had cut my finger with my Boy Scout knife. Two days before I left to return to San Francisco, I noticed that it was throbbing. Despite the fact that I was a physician's son, I'd never heard of infection and didn't bother to ask the camp doctor why the wound was turning yellow. When I got home, my grandfather Alvarez and my father asked me about my camp experience. I said that I'd had a wonderful time but that my finger was pretty sore. When I showed it to them, a look of absolute horror crossed their faces. They immediately took off my shirt and found the veins in my arm throbbing dark blue and the lymph glands in my armpit seriously swollen. My father lanced my finger and had me soak it for hours. It was a long time healing; Dad was convinced that one more day at camp would have killed me. Penicillin was a great discovery.

When we finished grammar school, in 1924, most of my classmates went on to a college preparatory school. Because I was keenly interested in mechanical things, my father sent me to Polytechnic High School instead; it offered vocational training to students not usually destined for college. As one of the few enrolled in an academic program at Poly, I was a fish out of water. I did learn mechanical and freehand drawing. I've used my mechanical-drawing training often, but only once did I use my sketching skill: when flying in a B–29 near Trinity Site, New Mexico, in 1945, when I realized that no one had thought to bring along a camera to record from the air the first nuclear mushroom cloud.

By 1925 my father had become the most sought-after internist in the Bay Area, an authority measurable by the number of hours I sat in his car outside hospitals where he gave other doctors advice. He reduced his private practice as his fees rose and spent more time at the Hooper Foundation pursuing physiological research. A member of the Mayo Clinic governing board heard him give a paper on his research at a medical conference. Soon thereafter the clinic offered him a full-time research position with no clinical responsibility, with paid assistants, and with a good salary. It was a fabulous opportunity, and he accepted. (When he retired from the Mayo Clinic, he became a syndicated and widely respected newspaper columnist. Many of his former readers remember him as "America's family doctor.")

Rochester, Minnesota, was a town of fifteen thousand with a single industry, the Mayo Clinic. In many ways it was like a college community, complete with divisions between town and gown. And it was cold. None of us had lived in cold country before. The winter, with temperatures not uncommonly forty below zero, was a jolt. We spent the day of our arrival, in February 1926, buying fur-lined coats and overshoes. I learned very quickly to ice-skate.

Our house was on the edge of town, far from the nearest bus line. We children walked a total of eight miles to and from school every day. Once when my sister Bernice and I were pushing our way through a fierce blizzard, she announced that she couldn't go on. She meant to lie down in the snow and sleep—and freeze. I needed all my young strength to lead and push her the last few hundred yards to home. After that first winter, when my father and I tried without much success to learn the intricacies of stoking and banking a coal-burning furnace, we moved to a house closer in with a furnace that regulated itself and burned oil.

Rochester High School was completely different from Poly. The quietly academic atmosphere was probably due to the number of doctors on the school board and to the townspeople's regard for the scholarly men who had brought the town international fame. But the most obvious difference between my life in Rochester and San Francisco was social. In high school in California I had never visited the home of a friend. With the exception of one party none of my friends had visited me. In Minnesota we were in and out of each other's houses day and night, girls and boys both. I skated every afternoon in a mixed group, played mixed-doubles tennis, and went to many dances. I felt like a country cousin at first, but since I had been to dancing school in California and knew how to dance my new friends regarded me as something of a city slicker. If I had remained in San Francisco, I think I would be a different person. In Rochester I came out of my shell.

My high-school science courses, although adequately taught, were not very interesting. I did find my first exposure to a professional physicist exciting. My father took me to hear an ophthalmic optics expert lecture on the electromagnetic spectrum, from gamma rays through visible light to radio waves. I wondered why he didn't put sound waves on his chart. Noting my interest in the lecture, my father suggested I might want to be a physicist someday. There were many physics books in his library, copiously annotated. When I asked him what physicists did, his first example was the new mechanical RCA phonograph.

With a scientific career a possibility for me, my father hired one of the Mayo Clinic's machinists to give me private weekend lessons. I worked in the clinic instrument shop during my junior and senior high-school summers. Some of the work was difficult as well as tedious, but I learned the tricks of the trade, including the seldom-practiced art of cutting gears. Because I did a lot of dirty work the first summer without complaining, the machinists went out of their way to teach me their skills during the second summer. I was a good pupil.

I acquired a characteristic in Rochester that has been important to my scientific career. I hesitate to recommend it to everyone. All the scientists I know are law-abiding citizens, but they have a healthy skepticism about authority. We are trained to ask "Why?" continually. At the same time we have to be judicious in questioning authority and its regulations. Someone who believes everything he is told simply can't be a scientist, but someone who believes nothing will wind up in jail or prematurely buried. In Rochester a friend and I used to climb the buildings under construction, usually by sneaking past the guards in the middle of the night. We climbed the three-hundred-foot clinic tower when it was only a skeleton of steel beams. We explored the power house and scaled the inside of its two-hundred-foot brick smokestack. I mention these escapades not to brag about being a scofflaw but only because I'm convinced that a controlled disrespect for authority is essential to a scientist. All the good experimental physicists I have known have had an intense curiosity that no Keep Out sign could mute. Physicists do, of course, show a healthy respect for High Voltage, Radiation, and Liquid Hydrogen signs. They are not reckless. I can think of only six who have been killed on the job.

The summer I graduated from high school, in 1928, my father took my brother, Bob, and me on the first Sierra Club High Trip arranged outside the United States. For a month we climbed mountains in the Tonquin Valley of the Canadian province of British Columbia.

Bob and I decided one day to climb a steep glacier that rose up behind our campsite. I had read Dad's mountaineering books and told Bob I'd show him how to climb. The glacier, covered with a foot or so of well-packed snow, sloped at about forty-five degrees; I led as we climbed across it, and only occasionally had to cut a step in the snow. The ice ended abruptly a few hundred feet downslope from our path and dropped off five stories to a rocky terminal moraine.

Far above us rose an impressive rank of palisades. At the beginning of our crossing, the ice and snow were smooth. Then we began seeing large pockmarks. I realized what caused those marks when I saw boulders melting loose from the palisades and cascading our way. One boulder hit a rock and began bouncing toward us. I yelled a warning to my brother and ran across the slope to escape the boulder as it bore down on us. I'd hardly started running when I hit a slick patch of ice. The next thing I knew I was sliding down the steep slope toward the edge of the glacier and the rocks below.

My armchair mountaineering came to the rescue. I remembered what I was supposed to do and did it: rolled over onto my stomach and jabbed the point of my pick into the ice. It worked; I stopped sliding.

Shaken, Bob and I quickly retreated from the avalanche chute we had stupidly crossed and made our way back to camp. When I returned to Rochester, I checked my father's books to see if I had handled the emergency properly. I quickly found the section that I had recalled so handily on the glacier and read through it. One paragraph stunned me. If you find yourself sliding down steep ice, it said, grasp the *head* of your ice ax with your free hand and use that hand to jab the pick into the ice. That was a vital detail I had forgotten; the book warned that if you simply hold the ax by its handle you risk losing it and finding yourself sliding on down the ice with nothing left to stop you.

I did my first roped rock climb under the tutelage of a Swiss guide later that week and subsequently climbed Mount Resplendent, my first climb through snow and ice. The young people in the party often hiked and ate together, thus encouraging my first summer romance. The trip served as a fine transition between high school and college.

I had always assumed that I would attend the University of California. My mother and most of my uncles and aunts had graduated from Berkeley, and my sister Gladys was enrolled there. But my Rochester teachers suggested that I consider the University of Chicago instead, stressing its strength in science. My father was enthusiastic about Chicago. All three American physics Nobel laureates had Chicago associations: Albert Michelson and Robert Millikan had done much of their prizewinning research at Chicago; Arthur Compton had just moved there. I applied and was accepted as a freshman in the class of 1932.

I was excited to be on the Chicago campus for the first time. My father came with me. It was one of those crisp, invigorating autumn days. Chicago weather usually leaves much to be desired, but that Saturday afternoon was splendid. We toured the campus admiring the ivy-covered Gothic buildings and exploring the interior of Ryerson Laboratory, which housed the physics department. I came to know every nook and cranny of Ryerson, although I wouldn't enter it again for more than a year. My father and I stopped at a number of research rooms on the first floor. In one we found a man setting up a new instrument. The instrument was an X-ray double-crystal spectrometer, and the man was Arthur Compton. The instant kinship scientists discover even across differences of discipline immediately bound the physiologist and the physicist. They were soon discussing Compton's new instrument and his research plans. For a man who had received a Nobel Prize only ten months ago, he was certainly not resting on his laurels. He was friendly to me, talked to me, and appeared to me to be quite old, though in fact he was thirty-six. I can't think of a more appropriate way for a freshman inclined to science to spend his first day on campus than in conversation with a Nobel laureate.

I went to the University of Chicago expecting to become a chemist. All the popular-science books I had read in high school praised the lives and great deeds of chemists. It's hard to realize how recently the word "physicist" came into common usage. In the 1930s, after I earned my physics Ph.D., I usually told laymen who asked that I was a chemist. Otherwise a long explanation would have been required.

Freshman physics, which I took in my sophomore year, I even found dull, a repetition of my high-school course. I was of two minds about organic chemistry, however. I loved the lectures, became adept at drawing the structural formulas of complicated organic compounds, and was able to predict accurately how compounds would change in reactions. But the chemical laboratories were repulsive. My one remaining memory of the qualitative-analysis laboratory is filled with hydrogen sulfide fumes. The organic-chemistry laboratories were even worse. Besides enduring the stenches of formaldehyde, benzene, and other pungent compounds, I never produced anything in my test tubes except a succession of black, gooey residues. I was unable to see any connection between the beautiful theory I had been taught in the morning and the sticky, smelly black masses I found in my glassware in the afternoon.

There were thirty fraternities on campus. I pledged Phi Gamma Delta, a house somewhere in the middle of the social spectrum. Fraternities have gone out of style at most of the better universities in the United States; only about half a dozen remain at Chicago, including mine. I lived in the fraternity during my undergraduate years and found my friends there. It was the center of my social life.

My sophomore year differed from my freshman in only one way: I fell in love for the first time. My closest upperclassman friend at the fraternity, Hugh Riddle, was engaged to Katherine Madison; Frances Madison was her younger sister. Hugh and Kay arranged a blind date for us, to a football game and the fraternity dance afterward. We dated each other exclusively for the next two years.

Hugh, Kay, Frances, and I spent many evenings aboard the Riddles's twenty-six-foot cabin cruiser, the *Lo-Ha-Lu,* berthed in Jackson Park harbor. On hot summer evenings the boat was delightfully cool. We cruised the shore south to the center of the city and then back northward, watching the city lights. For five summers, beginning in 1931, Hugh, three other fraternity brothers, and I explored Lake Michigan on two-week cruises. We were a close-knit and happy crew. Each of us had his own job. In the American tradition that the boy who owns the football is the captain of the team, Hugh skippered the *Lo-Ha-Lu.* Creighton Cunningham, who was working his way through college as a short-order cook, was our chef. Gordon Allen, a gossip columnist in the undergraduate newspaper, was historian and purser. Joe Bailey, studying law, served as steward, also known as bartender. I was the boat's navigator and carpenter and answered to the traditional nickname Chips. The warm friendships we formed on these cruises led to years of boat trip reunions.

I stuck it out as a chemistry major until the first quarter of my junior year. My final performance, in quantitative analysis, was important to my career as a scientist. The laboratory would soon close for the quarter, and I still had one unknown substance to analyze. I had divided my sample into three parts to analyze each independently and then average the results. I spoiled one sample. That left me with two and only a few days to produce an answer on the correctness of which my course grade depended almost entirely. If my result differed from the correct result by more than a no-go percentage, I would have to redo the analysis for very little credit.

My two answers differed by more than the allowable margin. I tried to think what I had done differently during the two procedures and concluded that I had lost some of the sample in the first run. I therefore arbitrarily raised my first assay to a value higher than the second, averaged the two, and submitted my answer. The laboratory assistant solemnly retrieved the coded standard and then burst into a big smile. I had hit it on the nose, he said, and would get an A for the work and a B for the course.

I had two reactions to this experience. First, I couldn't stand the idea of being a B chemist (I had managed seven straight B's in my chemistry courses) when it was beginning to be clear that I might be an A physicist. Second, I imagined with horror a lifetime in chemistry always guessing the answer. (For a long time I felt that I had violated scientific ethics by fudging my results. I later found that this process is allowed under the rubric "taking into account systematic errors," which makes it more palatable. It is in fact an important part of the professional scientist's way of life. Enrico Fermi was so much better an experimental physicist than most of the rest of us in part because he frequently changed a measured result after balancing all the factors that might have pushed the answer up or down.)

It really is difficult to make precise measurements. Everyone who has examined the way the best values of the fundamental physical constants have varied with time has noted that there is usually a "fashionable" value that is often many standard deviations away from a later value more precisely known. This phenomenon, which I call intellectual phase lock, occurs partly because no one likes to stand alone.

The person I know who most successfully avoided intellectual phase lock was Frank Dunnington, who worked with Ernest Lawrence at Berkeley in the 1930s. Dunnington spent several years measuring the electron's charge-to-mass ratio, e/m. He used a method that Lawrence had suggested shortly after he invented the cyclotron, his wonderful instrument for accelerating particles by spiraling them through a magnetic field: the method involves measuring the time it takes an electron to move through some measured fraction of one revolution in a magnetic field, a measurement from which the value of e/m can be calculated. The measurement was potentially much more accurate than earlier measurements using different techniques.

Dunnington knew that his e/m results were eagerly awaited but recognized his human susceptibility to intellectual phase lock. If he was going to spend four years of his life on a measurement, then he had to devise a scheme to avoid tilting the answer to an anticipated value.

He did so by deliberately obscuring a crucial piece of information, the angle between the slits in his experimental arrangement through which the electrons entered and exited. He told the head machinist that he wanted the angle somewhere between eighteen and twenty-two degrees, its precise value unknown. The machinist complied, and Dunnington began his years of careful measurements and the elimination of all "systematic errors." Only at the end, after he had written the final draft of his lengthy paper, did he dismantle his apparatus, remove the key slit system from its vacuum chamber, and mount it on a machine for measuring angles. After a long series of careful measurements, he wrote down the best measured value of the angle, punched that value into his desk calculator, and multiplied it by the number that represented the sum total of his several years of work. The number that appeared in the window of the calculating machine, now the world's best value of e/m, he wrote in the blank space he had left on the last page of his paper. Then he sealed the envelope and mailed it away.

Dunnington's care to avoid intellectual phase lock illustrates one major difference between scientists and most other people. Most people are concerned that someone might cheat them; the scientist is even more concerned that he might cheat himself.

I took six undergraduate mathematics courses. Until my junior year I encountered no one who seemed to have a greater aptitude for mathematics than mine. It had always been my easiest subject in high school, and I had been chided there by a classmate for writing my final exam with a fountain pen. Before I discovered physics, trigonometry and differential and integral calculus were my greatest intellectual pleasures. But my final mathematics course as an undergraduate was differential equations and the instructor gave me only a B. I'm sure he recognized that I was competent—I had worked every problem in the book—but he had to give the A's to the obviously brilliant students, who were now closing in on me. If I had decided to become a professional mathematician, as I easily could have, I would have made the traumatic discovery that there were many people my age who were far more talented mathematically than I could ever be.

The world of mathematics and theoretical physics is hierarchical. That was my first exposure to it. There's a limit beyond which one cannot progress. The differences between the limiting abilities of those on successively higher steps of the pyramid are enormous. I have not seen described anywhere the shock a talented man experiences when he finds, late in his academic life, that there are others enormously more talented than he. I have personally seen more tears shed by grown men and women over this discovery than I would have believed possible. Most of those men and women shift to fields where they can compete on more equal terms. The few who choose not to face reality have a difficult time.

At the YMCA in San Francisco I had learned simple gymnastic workouts on the horizontal and parallel bars, routines I practiced occasionally in Rochester. At the University of Chicago a member of the gymnastics team told me that Chicago had the best coach and usually the best team in the Big Ten. I stopped by the gymnasium one afternoon in my freshman year and was hooked. I spent two hours every day for the next four years practicing gymnastics, winning a varsity letter in my senior year. When I watch gymnastic competitions today, I'm struck by the routines, most of which had not been invented when I was a gymnast. The single most important characteristic of my success in physics has been invention. Whenever anything has interested me, I have instinctively tried to invent a new or better way of doing it. Why didn't I invent any new gymnastic routines? I'm astonished that I could have spent ten hours a week for four years doing and watching gymnastics and never try to improve it. Intellectually I was a different person then from the person I later became.

I discovered physics as a junior at the University of Chicago. It was love at first sight. My adviser enrolled me in a laboratory course, "Advanced Experimental Physics: Light," supervised by George Monk. Every scientist can recall the teacher who aroused his interest in a field. In my case it was Monk. He made it obvious to me that I had to be a physicist. I had been shielded from financial worries and didn't think to wonder how a physicist might earn a living.

The course laboratory, with equipment that Monk and others had painstakingly assembled over a period of years, occupied most of the

Ryerson Annex attic. On the center table was an optical spectrometer, a device for measuring the positions of spectral lines, the characteristic light signatures of elements. Its dispersing mechanism was a plane diffraction grating ruled with loving care in the Ryerson basement by Albert Michelson's optical technicians. Diffraction gratings, the precise instruments scientists use to split light into its component colors, are today pressed like phonograph records from a master. The grating in Ryerson Annex had 100,000 parallel grooves, each scribed mechanically onto a plate of optically polished metal by a carefully shaped diamond point. The grooves covered an area of about two by four inches, and they had to be spaced equally to an accuracy of about one-millionth of an inch. I marveled at the beauty of the grating and at the exquisite precision of the machine that made it, on which Michelson's team had worked for some thirty years. Each groove took five seconds to rule, the grating about a week, during which time the temperature of the ruling engine had to be carefully controlled. From beginning to end the diamond traveled three miles.

In my first experiment I used the Michelson diffraction-grating spectrometer to determine the wavelengths of the light from a mercury-vapor lamp. I was completely absorbed. Measuring wavelengths was the most exciting work I had ever done. Monk hovered in the background to answer questions and give advice. I was obviously fascinated with his course, and he devoted a great deal of time to me. My thorough appreciation of his teaching and his instruments must have pleased him.

The Michelson two-mirror interferometer, for measuring light interference phenomena, was my second-favorite apparatus. I had the privilege of using an instrument built in Michelson's shop and frequently adjusted by Michelson himself. (I never saw Michelson; he died in California that year.)

The joy I felt in working with this beautiful optical equipment, coupled with the trauma of quantitative analysis, made it clear to me and my adviser that my major should be physics. I had to make up for lost time by taking twelve physics courses in five quarters. Simultaneously I encountered the physics library. Until then my scientific education had come from textbooks, which are distillations and interpretations of original scientific papers. Now I discovered the originals themselves. Monk referred me to Michelson's work, much of which I found I could understand. Within a few months I had read every word he published in his long and distinguished career. He played an important part in my growth as a scientist; he was my first scientific hero.

The physics library was so engrossing that I had to force myself to leave it for food and friends. I'm convinced that a wide familiarity with the original physics literature is absolutely essential to a career in experimental physics. My excellent memory for material published in physics journals has been one of my most useful resources. I can't reproduce equations or text from memory, but I can often remember authors' names, which journal, and even the precise location of an important graph in an article read rapidly years before. My memory has come to be so extensively cross-referenced that I can walk into a library and find in a minute or two almost any article I've ever read.

My interest in physics having been aroused, I was no longer content to sit out the summer. I enrolled in three physics courses and finished my undergraduate work three months early. During a short vacation in Rochester I noticed the rainbow colors that reflect from a phonograph record and realized that the record was acting as a coarse diffraction grating. With a broken record, a light bulb, and a yardstick, I set about measuring the wavelength of light in our living room. When I showed my Chicago adviser my results, he suggested I write them up and send the manuscript to *School Science and Mathematics,* which I did. That was my first scientific paper; it was published in January 1932.

For my first research project in those last undergraduate months, my adviser suggested I build a Geiger-Müller counter (known as a Geiger counter now). No one in Chicago had seen one. The world literature on the subject consisted of only Geiger and Müller's original 1928 article, a few German articles, and three American articles that did not discuss construction details. So the project was both important and at the limit of my skills, a well-conceived assignment.

I was given a room of my own, which I learned to my great awe had been the site of Robert Millikan's historic oil-drop measurement of the charge of the electron. I had to build the metal parts of my counters in the student shop, seal them into glass envelopes, evacuate them with a vacuum pump I found in my research room, and learn the art of glassblowing along the way. The most difficult challenge was the amplifier. For the first two months of operation, I had no way of knowing if electrical breakdowns came from the counters or the amplifier; the physics department had only one cathode-ray oscilloscope, and as low man on the totem pole, I was never able to use it.

My Geiger counters had two distinctions: they were the first built in Chicago, and they were the worst counters ever constructed that actually functioned. They had a very high background of clicks even when no radioactive source was present. Nevertheless, I used them a year later to make a major discovery.

Robert Oppenheimer used to tell of the pioneer mysteries of building reliable Geiger counters that had low background noise. Among his friends, he said, there were two schools of thought. One school held firmly that the final step before one sealed off the Geiger tube was to peel a banana and wave the skin three times sharply to the left. The other school was equally confident that success would follow if one waved the banana peel twice to the left and then once smartly to the right. (My counters were unbelievably bad because I didn't use either of these techniques.)

In the course of time I was invited to demonstrate my counters at one of the weekly physics department colloquia. Since I was an undergraduate, I was not allotted a full hour; Arthur Dempster, who would later identify the historic uranium isotope U-235, used the first half of the hour to report on the newly discovered neutron.

Arthur Compton learned of my Geiger-counter work at that colloquium. He dropped into my laboratory later to propose that I upgrade my counters to make cosmic-ray measurements. The German physicist Walther Bothe had arranged two counters in such a way, I learned then, that no output click would occur unless both counters simultaneously detected the passage of a particle. This "coincidence technique," for which Bothe won the 1954 Nobel Prize, has two great advantages: it virtually eliminates the background noise, and it makes of the two counters a telescope capable of roughly determining the direction from which a particle has approached. No previous cosmic-ray detector had been directional.

I accepted Compton's suggestion and set out to build a Geiger-counter telescope. The work occupied my first summer in graduate school.

When I enrolled in graduate school in the spring of 1932, I changed my life-style significantly. I moved out of Phi Gamma Delta and into a graduate scientific fraternity, Gamma Alpha. Except for its Greek-letter name, Gamma Alpha had little in common with the undergraduate fraternities: no rituals, no regular Monday-night meetings, no membership committee. Anyone who wanted to join was welcome since every additional member reduced our share of the rent. Half of our members were studying physics; the others, geology, chemistry, and the biological sciences. We ate in-house to save money and socialized for a few minutes before and after meals. Bridge games sometimes convened, and the upright piano in the living room was usually under attack.

I renewed my interest in the piano and acquired a teacher, Frank Miller, a graduate student in physics who was an accomplished musician. He taught me the elements of harmony. Until then I had been able to play, rather badly, only classical compositions. I had dreamed of learning to play by ear but had heard from those who knew how that such an ability was a gift. Frank taught me otherwise. To play by ear, he said, I simply had to learn chords and progressions, no more difficult than learning integrals in mathematics. (I didn't learn then, although I managed to annoy any number of Gamma Alpha bridge players while practicing. I did finally acquire the skill in the early 1950s and discovered I had a good ear for harmony. Just as I found a more or less natural limit to my mathematical abilities that I could breach only with great effort, so I found I could play by ear any music I had ever heard—with hardly a thought—in the key of F; transposing into any other key required the application of great intellectual force.)

Fifteen graduate students crowded into two upper floors of the Gamma Alpha house. Five small rooms held our clothes and books; we shared one bathroom and a large, closely packed communal bedroom. These quarters, only slightly more confined than those in my undergraduate house had been, horrified my parents. I would now find it trying to live in such a pigsty, but Gamma Alpha allowed us to immerse ourselves totally in science. I spent most of my time with my fellows, a practical consequence of which was that I learned to benefit from the superior skills I found in my theoretically inclined friends and in those friends who worked in other areas of experimental physics. Constantly in and out of each other's research rooms, we shared what we knew.

Arthur Compton became my graduate adviser. He had just moved on from the X-ray work for which he won his Nobel Prize to an ambitious program of measuring the intensity of cosmic rays at widely spaced locations all over the world. (He was the ideal graduate adviser for me: he came into my research room only once during my graduate career and usually had no idea how I was spending my time.)

The nature of primary cosmic rays was not yet understood. Robert Millikan at the California Institute of Technology had seemed to show by a long series of experiments at various latitudes from Texas to northern Canada that cosmic rays were electrically neutral and therefore most probably a form of light—energetic gamma rays. He then wildly speculated that these gamma rays were emitted in cosmic space when particles combined to form atomic nuclei, and he popularized the notion that they were the "birth cries" of atoms.

In 1927 a Dutch physicist, Jacob Clay, had carried an electroscope with him on a voyage from Holland to Indonesia and found that the cosmic-ray intensity dropped by about 15 percent as he neared the equator. That would indicate that cosmic rays were charged particles caught up in the earth's magnetic field rather than gamma rays. But hardly anyone paid attention to Clay's results; they contradicted those of the famous Millikan.

In a year of world travel Compton confirmed the Dutchman's work. Millikan had limited his travel to higher latitudes, where the extra charged particles let in by the effectively weaker magnetic field were absorbed by the atmosphere. By traveling a wider range and by measuring at both high and low altitudes, Compton proved the existence of the latitude effect. Millikan did not accept Compton's results with grace.

But were the particles negatively or positively charged? During the 1932 Thanksgiving holiday the American Physical Society held its annual meeting at the University of Chicago. Attracted by the famous scientists who would attend, I stayed on through the holidays. Manuel Vallarta of the Massachusetts Institute of Technology discussed how to determine the sign of the cosmic-ray electric charge by using a pair of Geiger counters. He proposed measuring the cosmic-ray intensity at an appropriate elevation angle first over the eastern horizon and then over the western. Oppositely charged particles should be deflected by the earth's magnetic field in opposite directions. If more cosmic rays came from the east, they were negatively charged, as expected; if more came from the west, they were positively charged.

The experiment had been tried before, inconclusively, particularly by Bruno Rossi of Florence. Vallarta pointed out with great excitement that earlier researchers had done their looking in temperate latitudes, where the magnetic field had no measurable effect. He predicted a large effect in his native Mexico City, which was favorably located not only in latitude but also in altitude. He promised to host any physicists willing to conduct this important experiment there.

Arthur Compton had no apparatus of his own that was directionally sensitive. He asked me if I would like to take my Geiger counters to Mexico City. Tom Johnson, a cosmic-ray physicist at Swarthmore with an operational cosmic-ray telescope, also accepted Vallarta's invitation.

I worked around the clock preparing my equipment. Vallarta warned me that the Mexican electrical supply was notoriously unreliable. I converted my counters to run off several dozen heavy forty-five-volt batteries and put together a kit of hand tools and spare parts. At Compton's suggestion I took my apparatus to Chicago's Jackson Park and operated it successfully for two days. That gave me the confidence I needed to embark on a physics expedition to a strange land.

The Johnsons and I arrived in Mexico City the same day. Vallarta met us at the station, shepherded us through customs, and drove us to the Geneve Hotel. He arranged for riggers to hoist our equipment to the hotel roof, which we reached by ladder from the top floor, where we took rooms. Manuel also found some canvas for me, from which I fashioned a makeshift tent. Tom had brought along a handsome commercially made tent to house his two beautiful sets of apparatus, beside which mine looked exceedingly crude. His telescopes, each with three Geiger counters, rotated about a vertical axis, so that he could line them up facing east or west at any elevation angle. He kindly pointed out that mine needed this feature. I had mounted my two counters on the hinged wooden lid of my battery box and could set them to any necessary elevation angle simply by raising the lid. Tom thought I should also rotate the tubes to avoid introducing a possible bias. For a rough-and-ready remedy I bought a wheelbarrow, which I hauled up the side of the hotel with a rope. I placed my detector with its battery box on the wheelbarrow and changed from east measurements to west every half an hour by moving the wheelbarrow around.

After a few days of testing, Tom and I independently began to see a higher counting rate coming from the west than from the east. We celebrated at dinner that night, confident we had shown primary cosmic rays to be positively charged—most physicists had guessed the opposite. We thought they might be positive electrons, positrons, discovered the preceding year. (They turned out to be a representative sample of all the nuclei in the universe, with protons forming the largest proportion and energetic gamma rays accounting for fewer than one in ten thousand.)

My stay in Mexico City lasted a month. It was extraordinarily pleasant. My fussy apparatus kept me busy. I studied physics texts. I was invited to a number of parties attended by beautiful girls my own age who had been educated in the United States and spoke perfect English but who were undatably well chaperoned. I had previously known only bootleg gin; I sampled aged Scotch and after-dinner liqueurs with real pleasure. The Johnsons were surprised to find me so young and so inexperienced in the world of physicists. Tom had been a graduate student at Yale University with Ernest Lawrence. I learned much from him about the personal side of the man who would later become my mentor.

When I returned to Chicago, Compton and I went over my data and agreed that the experiment had been satisfactorily performed. We then collaborated in writing a letter for the *Physical Review.* By prior agreement it would be published in the same issue with Tom Johnson's report. Arthur Compton's name came before mine in the signature, and the name order seemed quite reasonable to me. A few days after we mailed our letter, Compton learned that Tom planned to discuss his observation of the east-west effect at a physics meeting. Compton felt that such public announcement prior to the publication of our letters violated the spirit of our agreement. His competitive sense aroused, he arranged to present our results at a meeting of the National Academy of Sciences scheduled a few days earlier than Johnson's. I was surprised and pleased when Compton told me subsequently that he had telegraphed the editor of the *Physical Review* to change the order of our names. He said that we should share the credit equally and that his name had come first on the academy report (to which I have never seen a reference). Unexpectedly I became first author of a widely referenced paper that described a significant discovery in basic physics.

The Century of Progress Exposition opened in Chicago in 1933. Its theme was scientific progress. The previous Chicago world's fair had been held forty years earlier; light from the star Arcturus, forty light-years away, signaled the start of the 1933 exposition, symbolically connecting the two events.

I similarly supplied a dressed-up Geiger-counter telescope to signal the start of a Chevrolet assembly-line exhibit. Alfred P. Sloan, the chairman of General Motors, had called Arthur Compton before the exposition opened and asked him if he knew of some celestial phenomenon more glamorous than forty-year-old starlight that could be harnessed to open the GM exhibit. Arthur nominated cosmic rays, which he said had been traveling through the universe for a million years or more. He also nominated me to build a suitably picturesque detector.

With Les Johnson, who had just left the university to begin a custom radio business, I spent night and day working to assemble the appropriate equipment in time for the opening banquet. Sloan would be sitting at the center of the bend of a U-shaped banquet table. We planned to set up our cosmic-ray Geiger-counter telescope on view directly in front of him. The telescope would be connected to two large neon signs; whenever a cosmic ray passed through the counters, the neon signs would flash COSMIC RAY. Sloan would activate the demonstration assembly line at some appropriate point in his speech by pressing a gold-plated telegraph key. The key would shunt the next signal that a cosmic ray induced through a series of increasingly large electric relays, finally starting the electric motors that drove the equipment on the production line.

We had the telescope display case built out of painted plywood and silver-plated the glass cylinders that housed the Geiger counters. When we set up on the morning of the banquet, everyone found the equipment handsome. We had no time to run reliability tests, so we took the precaution of stringing an extra set of wires to the power supply for the neon signs; if the telescope failed, one of us could stand by and click the wires together unobtrusively to simulate cosmic-ray detection.

Activating the production line was spectacular. We didn't need our backup wires. Sloan held the golden telegraph key aloft for everyone to see as he pressed it. A few seconds later a huge din followed the flash of the large neon COSMIC RAY signs as the automobile workers began banging on the bodies of an advancing line of Chevrolets. It was my first experience with industrial consulting, and Arthur Compton saw to it that I was well paid. He proposed that GM send $1,250, kept $250 for a finder's fee, and gave me the rest. I paid for our materials and split the remainder with Leslie; we each got $425, far more than we would have requested. For perspective on our windfall, consider that three years later, with a Ph.D. and a wife, I accepted an offer of $1,000 for a full year's work.

Remembering that county fairs in those days usually featured balloon flights, the directors of the exposition wanted to sponsor a suitably spectacular balloon flight that would try for a new world altitude record. Compton was overjoyed at the opportunity to fly cosmic-ray apparatus in the gondola. I worked hard building the apparatus, installing it in the gondola, and testing its reliability. On one occasion when I was in the metal gondola working with the 1,500-volt cable that connected to the batteries under the floor, I must have imagined that I was back in Ryerson Laboratory, where the floor was made of wood and the lab table of marble, both nonconductors. There I had become used to picking up high-voltage wires with no fear of shock —I wasn't grounded. I did the same thing in the gondola and received a very nasty shock that could easily have been fatal. I let out a loud curse of the kind favored in fraternity circles and glanced up to see Arthur Compton looking at me through the hatch. He taught Sunday school every week; I'm sure he hadn't heard language like that in years.

The pilot was to be Navy Lieutenant Commander "Tex" Settle, a famous free-balloon pilot who had won several international races. Jean Piccard, the current record holder's identical twin brother and an organic chemist, was supposed to serve as copilot and scientific observer. Probably because Piccard and Tex Settle didn't hit it off, Piccard was removed and I was offered his place.

I discovered then what it was like to be a celebrity. Pathé News wanted to film an interview with me for distribution to movie theaters throughout the United States, the only place Americans could see the news in those days before television. I knew that the medical profession frowned upon personal publicity, and I had some hints that physicists felt the same way. Around the laboratory I had heard that publicity was measured in an absolute unit, the "kan." That unit was too large for ordinary application and a practical unit one one-thousandth that size served in its place, the "millikan." In fairness, it must be said that Millikan used publicity as an acceptable and important tool to build Caltech into a great institution. (I never was filmed by Pathé News.)

I spent my first weeks as scientific observer with Tex Settle learning the lore of free ballooning. The emergency procedure required waiting until the gondola fell to 15,000 feet, getting out of the hatch in a hurry, snapping on a parachute, then free-falling while steering away from the gondola to avoid hitting it when the parachute opened. Tex thought this was a large order for a first jump and suggested I get some practice. He arranged parachute instruction for me at Great Lakes Naval Air Station. I went to sleep that night dreaming about jumping out of a Navy plane.

The next day the cloud on which I had been floating for the past several weeks gave way. Exposition lawyers concerned about Jean Piccard's response to my appointment vetoed my flight. I missed the parachute jump as well as the later flight. Tex Settle took off by himself from the center of Soldier Field as one hundred thousand people watched from the stands. (I was positioned directly underneath the balloon as it rose, majestically.) Unfortunately, because of a design flaw, the balloon went up only a thousand feet and then descended into a nearby railroad yard. I greeted Tex there a few moments later. He was as disappointed as I had been a few weeks earlier.

Once the gondola was repaired, the flight was rescheduled to leave from the great airship hangar at Akron, Ohio, and Major Chester Fordney of the U.S. Marines rounded out the crew. The one light touch in the whole affair came just before the balloon took off from Akron. One of the experiments aboard was designed to measure the effects of intense cosmic rays on fruit flies. The word got around that the sex of the flies might be changed; the reporters inflated that notion to suggest that the sex of the balloonists might be changed. The last object passed to the crew was a box of sanitary napkins, which proved to be among the most useful items aboard, since the pads served to soak up condensation on the gondola walls that might otherwise have dripped into the electronic equipment. The Settle-Fordney flight rose to 61,237 feet, a world record then. But for a few minor glitches, my curriculum vitae might now include the entry "1933–36, coholder of the free-balloon altitude record."

For Christmas that year my grandfather gave me his customary fifty-dollar check. Since I had lost my chance to parachute from a Navy plane, I decided to spend the fifty dollars on flying lessons. I drove out to Midway Airport and found a willing instructor at fifteen dollars an hour, a fee that included the use of a large, open-cockpit Curtiss Fledgling biplane, a direct descendant of the famous Curtiss Jenny of the First World War. I asked my instructor whether I would be able to solo on my fifty dollars. He said I should if I had any aptitude. I soloed with just three hours of dual instruction and made three solo landings before I used up my Christmas check. This early flying experience began my lifelong love affair with aviation. When

I closed my log books at the age of seventy-three, they showed more than one thousand hours of flying time, most of it as pilot in command. I think of myself as having had two separate careers, one in science and one in aviation. I've found the two almost equally rewarding.

That summer I began to feel that I was progressing toward membership in good standing in the scientific fraternity. The American Association for the Advancement of Science met in extraordinary session during the opening days of the Century of Progress Exposition. I attended all the physics meetings and was excited that I could sit through several days of scientific lectures and understand most of what was discussed. I heard Ernest Lawrence, John Cockcroft, Kenneth Bainbridge, Harold Urey, and Niels Bohr (except that Bohr's famously whispered words were inaudible from where I sat). I was reading nuclear physics by then, and for me Lawrence was the star of the AAAS meeting. His 27-inch cyclotron had accelerated its first heavy-hydrogen nuclei, deuterons, less than three months earlier, and every experiment that he and M. Stanley Livingston tried with their energetic deuteron beam was interesting and well beyond the reach of any other physics laboratory in the world.

Lawrence returned to Chicago just as the summer quarter was ending. My older sister Gladys worked part-time for him in Berkeley as a secretary and had mentioned me to him. He invited me to tour the exposition. I was dumbfounded and delighted. I still knew my professors only at a distance; socializing with a first-rank physicist was a revelation. I was unprepared for Lawrence's enthusiasm. He wanted to listen to the symphony for a while, he wanted to stop in one of the bars, he wanted to walk the exposition's Streets of Paris. He seemed to be a regular guy. He was then thirty-two, but I didn't feel any age difference between us at all. We walked at a good clip, and when we were finished touring we hiked all the way back to Lawrence's hotel. "Well," he said there, "let's have a nightcap." So we wandered into the bar and had a drink together. It was the beginning of a long and fruitful friendship.

TWO

Becoming a Physicist

IN his relatively short life of fifty-seven years, Ernest Lawrence accomplished more than one might believe possible in a lifetime twice as long. A South Dakota native with a doctorate from Yale, he left a Yale assistant professorship in 1928 for the University of California at Berkeley, where, two years later, he became the youngest full professor in the history of the university. It's difficult today to appreciate the courage Lawrence needed to leave the security of a rich and distinguished university and move to what was, by contrast, a small and only recently awakened physics department. In later life he would recall the universally dire predictions of his eastern friends; they agreed that his future was bright if he stayed at Yale but that he would quickly go to seed in the "unscientific climate of the West."

What happened, of course, was just the opposite. Lawrence did significant work in several fields of physics. But he is best and properly known first of all as the inventor of the cyclotron, the work for which he earned, in 1939, the Nobel Prize in physics.

In the period when Lawrence was moving from New Haven to Berkeley, physicists were excited by the news of the nuclear transformations achieved in Ernest Rutherford's Cavendish Laboratory, at Cambridge University in England. It was generally recognized that an important segment of the future of physics lay in the study of nuclear reactions, but Rutherford's tedious technique (using alpha particles from radium to bombard and perturb the atomic nucleus) repelled most prospective nuclear physicists. Simple calculations showed that one microampere of electrically accelerated light nuclei would be more valuable than the world's total supply of radium—if the nuclear particles (protons) had energies in the neighborhood of a million electron volts. As a result of such calculations, several teams of physicists set about to produce beams of million-volt protons with which to bombard the nucleus.

Lawrence had spent enough time studying spark discharges, akin to lightning, to develop a healthy respect for spark breakdown as a voltage limiter. Although he wanted to get into the nuclear business, the avenues then available didn't appeal to him, because they all involved high voltages and consequent spark breakdown.

In his early bachelor days at Berkeley, Lawrence spent many of his evenings in the library reading widely, both professionally and for recreation. Although he had passed his French and German requirements for the doctorate by the slimmest of margins, and consequently had almost no facility with either language, night after night he faithfully leafed through the back issues of the foreign periodicals. One night in the spring of 1929, browsing through a journal seldom consulted by physicists, he came across an article by a Norwegian engineer, Rolf Wideröe, entitled "Über ein neues Prinzip zur Herstellung hoher Spannungen" (On a New Principle for the Production of High Voltages).

Lawrence was excited by the easily understood title and immediately looked at the illustrations. One showed the arrangement Wideröe had employed to accelerate potassium ions to 50,000 electron volts through a hollow drift tube attached to a radio-frequency source of 25,000 volts. The Norwegian's innovation was the use of a relatively small voltage to produce increasing acceleration by alternating the voltage to push and pull the charged particles through the evacuated tube. Lawrence immediately sensed the importance of the idea and decided to try its obvious extension to many accelerations through many drift tubes.

"Without looking at the article further," Lawrence recalled in his Nobel lecture, "I then and there made estimates of the general features of a linear accelerator for protons in the energy range above one million electron volts. Simple calculations showed that the accelerator tube would be some meters in length, which at that time seemed rather awkwardly long for laboratory purposes." Since he could do his own thinking faster than he could translate Wideröe's paper, Lawrence had the pleasure then of arriving independently at many of the Norwegian's conclusions. It struck him almost immediately that one might "wind up" a linear accelerator into a spiral accelerator by setting it in a magnetic field. He was prepared to arrange the magnetic field to vary in some manner with the radius, so that the time of revolution of an ion would remain constant as its orbit increased in radius with increasing acceleration. But he found by calculation, on the spot, that no radial variation of the magnetic field was needed—ions in a constant magnetic field circulate with constant frequency, regardless of their radius. That meant slow ions and fast could be pushed and pulled in their smaller and larger orbits at the same rate and be accelerated at the same time. The slower ones were bent into circles with smaller circumferences, so the time per revolution was the same at all speeds.

In 1954, sitting in the Lawrence living room, I heard the fourteen-year-old Robert Lawrence tell his dad that he had been asked by his physics teacher to explain the cyclotron to his class. After Ernest had explained how the slow ions went around in smaller circles and the fast ones in larger, making it possible to push and pull all of them at the same rate, Robert said, "Gee, Daddy, that's neat." I related this incident in the biographical memoir of Ernest's life that I wrote for the National Academy of Sciences and added, "That is probably what the members of the Nobel Committee thought when they voted Ernest the Nobel Prize in 1939."

Lawrence and N. E. Edlefson reported the first demonstration of this resonance principle in the fall of 1930. The cyclotron, as it evolved, had two hollow brass electrodes shaped like the cut halves of a cylindrical pillbox, called "dees" because of their capital-*D* shape, mounted in a vacuum tank between the pole faces of a large electromagnet. The two dees, alternately charged, pushed and pulled ions injected at the center as the ions spiraled around in the magnetic field. After perhaps a hundred orbits the ions exited in a beam that could then be directed onto a target. By 1932 Lawrence and M. Stanley Livingston had built an 11-inch machine (the measurement referring to the diameter of the pole pieces—the dees were slightly smaller) and with high-frequency oscillations of only four thousand volts achieved energies of more than one million volts. When I heard

Lawrence speak at the Chicago exposition he had advanced to a larger, 27-inch machine.

Toward the end of the summer of 1934 my parents invited me to drive with them to California. I accepted the invitation partly to see Ernest Lawrence and his laboratory.

The first cyclotrons had been built and operated in LeConte Hall, which housed the Berkeley physics department. As soon as the 11-inch machine was operational, Lawrence had begun looking for funds to build a larger one. The most expensive part was the large electromagnet. The Berkeley physicist learned that two magnets larger than any he had hoped to find were available. Four had been built by the Federal Telegraph Company for the U.S. Navy to generate radio signals by a method that had subsequently become obsolete. The French bought two of the four, but the other two were surplus. Leonard Fuller, a Berkeley professor of electrical engineering who was also a vice-president at FTC, gave one to Lawrence.

Lawrence then persuaded President Robert Gordon Sproul to let him use an old wooden building near LeConte Hall that had belonged to the engineering department. There he had the huge magnet installed on concrete piers. With his wonderful intuition in experimenting, he redesigned the pole pieces and the copper winding to make them symmetrical above and below the magnet's central plane. Had he not done so, the 27-inch cyclotron would never have worked. He had a bit of luck—he picked, by chance, the better of the two surplus magnets. The other one, which later went to John R. Dunning at Columbia University, carried an internal air pocket in its pole piece that spoiled its field. If Lawrence had chosen that one, he might well have concluded that a large cyclotron couldn't be made to work, and nuclear physics would have been dealt a serious setback.

From the outside the old wooden Radiation Laboratory was unimpressive. Inside, it was the most exciting place I had ever seen. In only a few years Lawrence and his coworkers had built a cyclotron that accelerated deuterons to five million electron volts (MeV); two Wideröe linear accelerators that produced mercury and lithium ions of several MeV; and a Sloan resonant transformer that produced enormous currents of 1 MeV deuterons. I spent several days there becoming familiar with many new experimental techniques.

The Radiation Laboratory impressed me as much for its atmosphere as for its marvelous experimental equipment. Ernest established the style, and one felt his presence even when he was absent.

At Chicago we graduate students were close friends; we were interested in each other's work; we enjoyed a fine camaraderie in the halls. But it was considered a serious breach of etiquette for anyone to suggest how a friend's experiment might be improved. By contrast, everyone at the Radiation Laboratory was encouraged to offer constructive criticism of the experiments his colleagues were performing.

Sociologically, then, Chicago and Berkeley differed in their boundaries between private and community property. At Berkeley everyone shared. The Radiation Laboratory had no interior doors. Its central focus was the cyclotron, on which everyone worked and which belonged to everyone equally (though perhaps more than equally to Ernest). No one hoarded scarce commodities—the Chicago tradition—and there was a minimum of private property. Malcolm Henderson had "his" linear amplifier and ionization chamber, Ed McMillan "his" quartz-fiber electroscope, and Franz Kurie "his" cloud chamber, but everyone was free to borrow or use everyone else's equipment or, more commonly, to plan a joint experiment with the "owner." Ernest Lawrence's greatest invention was doing physics in cooperative teams. I was enormously stimulated by it.

One evening during our visit Lawrence invited my parents and me to dinner at his home. There I first met Molly, his wife, who would soon present Ernest with his first son. Molly, whom everyone who knew Ernest admired and loved, brought great warmth and stability to her husband's hectic, exhausting life. She truncated a promising career as a bacteriologist when she married Ernest, but she could have had a greater effect on scientific progress in this century only if she had been the Madame Curie of bacteriology.

When I returned to Chicago in the fall, I spent several months working on an experiment I had planned that involved the use of a very coarse diffraction grating. It was useful—I learned a lot from it—but Arthur Compton could have suggested more profitable avenues for me to follow. He encouraged me instead to think up my own problems. After I had finished the experiment, I went to him and told him that I had played around long enough with my own ideas and would like him to suggest a worthwhile doctoral thesis. He laughed and said that I should turn in my diffraction-grating experiment. I protested, but he laughed again and said it mattered very little what sort of thesis anyone submitted. If a man went on to teach in a small college somewhere, nobody would care about his doctoral thesis; the only important fact would be the degree itself. Nor would anybody care about thesis work if a man went on to become a successful researcher; people would be interested only in his current work. Eventually my thesis was published in the *Journal of the Optical Society of America;* that constituted the decent burial it deserved.

For my diffraction-grating experiment I used a very bright mercury-vapor lamp, which I built with the help of a friend who was studying chemistry. In use my lamp would blow up and spray mercury droplets all over my basement room on the average of once a day. Mercury in the cracks in the wooden floors of physics research rooms was not then considered a hazard but was accepted like ants at a picnic. My room was small, and its door, like that of a walk-in shop refrigerator, tightly sealed it. Books on industrial safety indicate to me now that I should have developed hatter's symptoms during my countless hours of exposure to mercury vapor. The mad hatters Lewis Carroll immortalized went mad from brain damage caused by breathing mercury fumes from liquid mercury used in the felt-forming process. (On the other hand, some of my colleagues may feel this episode explains my subsequent behavior very well.)

At this time I began to date Geraldine Smithwick, whom I had met when she was dating my friend Gordon Allen. Frances Madison and I had often double-dated with Gordon and Gerry; several months after they had broken up, I started taking Gerry to fraternity dances and other university parties. She was a senior, very pretty and very popular. In Robert Hutchins's first year as president of the University of Chicago, she was one of the dozen freshmen he and Mortimer Adler invited to take the famous "Great Books" course. She was active in student government and eventually became vice-president of the Associated Students, a position usually won by a man. She was president of the women's drama association and in 1934 was selected to lead the Interfraternity Ball. I attended the ball as her escort. At ten o'clock at night I walked in the ballroom procession in white tie and tails; at five o'clock the next morning, wearing an old suit and scuffed shoes, I boarded a Greyhound bus for a two-day ride to the annual meeting of the American Physical Society in New York, where I gave my first ten-minute talk.

In the summer of 1935 my Spanish grandfather wrote unexpectedly to propose that he send me to Europe for a year to study. I was overwhelmed by his generosity. Reading scientific articles in German was suddenly no longer a chore.

When I finished my diffraction experiment, I holed up in my office surrounded by a big stack of physics texts, determined to atone for my misspent youth. I had to learn physics theory to pass my doctoral oral examinations and had to do so from textbooks rather than from the several years of lectures I had neglected to attend. I was pleasantly surprised to find that I enjoyed the work. I had seen enough physics and physicists to realize that there were many interesting ideas of which I knew very little. Thus motivated, I studied the theory of the science that I now loved but really didn't understand very well.

By then Geraldine and I were engaged to be married. We had decided that we would marry in mid-April 1936, just after my orals, when I could properly call myself a doctor of philosophy. We looked forward to spending our first married year in Europe. We studied the deck plans of transatlantic ocean liners and were ready to make reservations. Then the roof fell in. My grandfather wrote that he now thought it was the wrong time for Americans to study in Europe: the civil war in his native Spain might spread. The wedding invitations were being engraved, I would soon be responsible for a wife, and abruptly I had no way to support her. (Gerry had a fine job as secretary to the president of a foundation, but it never occurred to us that she might support us; we had both been brought up in the old-fashioned tradition that the husband is the breadwinner.)

I wrote to a number of universities looking for work. I applied for a National Research Council Fellowship. Ernest Lawrence wrote in support of my NRC application, saying that I would be welcome in his laboratory. By the beginning of April nothing had turned up. I swallowed my pride and asked my sister if she knew any way I might get a job at the Radiation Laboratory. A few days later this wonderful telegram arrived:

> Dr. Luis Alvarez. Lawrence says come on out. If you get no fellowship can pay you probably a thousand a year depending on how much money he can get for budget. Can't start pay until July, but plenty of work any time you come. He sends congratulations for wedding. My letter follows. Gladys.

Gerry typed my thesis in her office while I studied for my orals. A few days after I passed them, Gerry and I were married in one of the university chapels. The next day we left for our new life in California. The four and a half years that followed offered the most concentrated and thoroughly rewarding experience of my career. I don't see how

I managed to accomplish so much in such a short time. I adhered completely to the Radiation Laboratory tradition that the most important thing in the world was physics; you were a dilettante if you didn't work at it for at least eighty hours a week.

I was still unprepared for the great adventure on which I was embarking when I rang the bell at the old wooden Radiation Laboratory that first May morning in 1936. A graduate student, Ernie Lyman, opened the door. Lawrence and his associate director, Don Cooksey, were away, Ernie said; he suggested I report instead to Jack Livingood, who was acting on Lawrence's behalf. Jack remembered me from my previous visit. When could I begin work? he asked. "As soon as I can get my coat off," I told him. That was good, he said, because there was a lot to be done. But Ernie insisted on first showing me around.

The laboratory had changed in the twenty months since my earlier visit. The cyclotron control table had been moved into an adjoining room shared with a drafting table and Don Cooksey's private workbench. (Each member of the laboratory staff had a few feet of workbench on which to keep personal possessions, but no one had a telephone or a desk. We were free to borrow from one another, so long as we left a note. Don was a connoisseur of fine tools and had the financial resources to acquire them; he proved to be a rich source of needed tools.) The control table supported panels of meters plus a few movable controls.

On start-up the cyclotron beam current was so small that it could be observed only on a wall galvanometer that projected a spot of light onto a translucent glass scale; it was a great day when the beam intensified enough to be read on an ordinary microammeter. The cyclotron operator maximized the beam current by pushing a movable slide back and forth on a wire-wound rheostat that adjusted the magnet current, and thereby the magnetic field, to match the driving frequency of the radio-frequency oscillator. When the beam current decreased, the operator had to decide in which direction to move his rheostat control. To make matters worse, the magnet's enormous inductance gave the system a response time of many seconds. Each operator learned by trial and error to attribute beam-current falloff to a magnetic-field overshoot, an oscillator-frequency drift, or a deuterium-pressure change in the vacuum chamber. Everyone prided himself on his skill as an operator, but Ernest Lawrence was still the best although he spent less and less time at the controls. Molly Lawrence was reputed to be a close second; she had perfected her skill during the long evenings of working with her busy husband. Eric Lawrence, two years old, sometimes came to the Radiation Laboratory with his father on Sunday morning. On one visit someone asked Eric how he was, and he responded, "I'm fine. How's the vacuum?" It was a reasonable question; the state of the cyclotron vacuum was a matter of hourly concern.

A laboratory visitor first passed through the machine shop and then into the control room. There he found himself face to face with the 27-inch cyclotron. Its distinctive magnet was as well known to nuclear physicists, and as much revered, as Mecca's Kaaba is to Moslems. The roots of modern high-energy physics run directly back to that cyclotron and those who worked with it.

The cyclotron had acquired a new vacuum tank since my last visit. Don Cooksey designed it. Don and Ernest had been good friends as fellow graduate students at Yale; when Don received his Ph.D., in 1934, Ernest invited him to join his cyclotron team. At the time Don arrived, the cyclotron vacuum chamber had evolved with no attention to sound engineering design. Its thick layer of wax bore witness to the constant battle waged with that part of nature which is said to abhor a vacuum. Don designed a sturdy new chamber. Perfecting his design over long hours at his drafting board, he introduced water and air cooling to keep things from melting. Large-diameter glass pipes made insulators to support the cantilevered dees. A mixture of beeswax and resin served as a vacuum sealant except where the metal or glass could be properly heated; then traditional red sealing wax was used. The resulting chamber was beautiful. As I came to understand every detail of its construction, I also came to admire the ingenuity and the loving care that Don had lavished on it. He was the first great instrument designer I was to know and also a man universally loved by everyone who knew him.

Ernie Lyman took me around to the back of the cyclotron, where a new oscillator had been installed. Oscillators were mysterious to me then. I realized that the most obvious lacuna in my technical knowledge was radio engineering, and on my way home that first evening I bought a text on the subject, which I studied thoroughly in the weeks to come.

The motor generator set that powered the cyclotron occupied an open court way. I would need days of experience before I learned to feel comfortable turning it on. Directly above it was mounted a weird contraption that demonstrated the ingenuity of Lawrence's team. The high-powered oscillator had to be started at a lower power and therefore needed a control knob. The standard control today is a transformer with a continuously variable ratio. Equivalent devices then were too expensive; the laboratory solved the problem by building homemade variable resistors in the form of flowing columns of water enclosed in six-inch glass tubes four feet tall with large, circular graphite electrodes at each end. The upper electrodes were connected together with rope over pulleys; all of them could be moved by turning a large wheel on the control table. Water flowed through the tubes and spilled out, an extremely wasteful arrangement. The laboratory wasn't charged for such services, however, and the university didn't object; physicists had not yet learned cost accounting.

Continuing the tour, Ernie led me back through the cyclotron room and into a mirror image of the control room that housed a linear accelerator. I was overwhelmed there by the nauseating odor of hundreds of caged rats. They belonged to John Lawrence, Ernest's brother, a physician who was investigating the biological effects of cyclotron radiations. Ernie worked in that room with his wife, a spectroscopy student, and I found it difficult to understand how they could bear the rat stench. Like them, I soon became immune and set up my first workbench a few feet from Ernie's.

The one small enclave on the ground floor of the building that had not yet been claimed by the Radiation Laboratory belonged to the chemistry department. Two graduate students, Glenn Seaborg and Dave Graham, had built a low-voltage deuteron accelerator there to produce neutrons by reacting deuterium with deuterium. (Glenn Seaborg has played several important roles in my scientific career, not the least of which was convincing the Nobel committee that I should be awarded the 1968 physics prize.)

As I walked back with Ernie to the cyclotron room, I took off my coat, rolled up my sleeves, and told Jack Livingood that I was ready to go to work. He said that a transformer that supplied the deflecting voltage to extract the deuteron beam had just burned out. Jack and I lifted the heavy homemade transformer down from its perch atop the magnet. Following his directions, I emptied it of insulating oil, a messy operation. Eventually the transformer core and its primary and secondary windings rested exposed on a bench nested on a big rag that soaked up the oil that continued to drip from its innards.

Watching Jack diagnose the trouble, I felt the same admiration a student of surgery would feel watching the chief surgeon cope with a medical emergency. Jack quickly deduced that the problem was caused by a backup of radio-frequency (RF) power. I unwound the charred secondary winding and put in a couple of electronic filters. We rewound the secondary, remounted it on the magnet yoke, filled its glass case with oil, and found it to be as good as new.

If a similar transformer problem had developed twenty-five years later, I would have called the engineers in electrical maintenance. They would have replaced the damaged transformer with a new one. In 1936 we had not one professional technician in service, let alone replacement spares.

There was so much radio-frequency power floating around the laboratory in those days that I could hold a hundred-watt light bulb by its glass envelope, touch its metal base to almost any electrical conduit in the cyclotron room, and make it light up, my body acting as an antenna. I would soon begin a relentless battle against these powerful RF fields.

The transformer repaired, I went home for lunch to present Gerry with her first whiff of the fumes of transformer oil that would signal my whereabouts for years to come. The cyclotron magnet's copper windings were contained in and cooled by two tanks of transformer oil that circulated through a heat exchanger. We frequently knelt on the floor and rested our elbows on the lower tank when we were searching for leaks in the cyclotron vacuum tank. Oil vapor from the hot liquid seeped up then and permeated our clothing and our hair. Like Li'l Abner in Al Capp's comic strip, we served periodically as inside men at the Skunk Works. "Oh, you must work at the Radiation Laboratory," strangers deduced about me immediately upon introduction at cocktail parties. Their evidence was strictly olfactory.

After lunch we turned on the cyclotron and I started my new life as an operator trainee. I was eager to learn; the crews appreciated my enthusiasm and soon automatically invited me to observe every operation they performed.

I stood behind Jack Livingood at the control table and watched him tap the face of the microammeter that indicated the state of the vacuum. Although it seldom does any good, everyone taps vacuum gauges, hoping that the needle is stuck. When Jack was convinced that the pressure would stay high, he announced that we would have to "flame the wax." We restored the vacuum by waving a Bunsen burner flame across the beeswax-resin mixture that sealed the vacuum tank, melting and resealing the outer layer. Just as California is crisscrossed with fault lines that can open to earthquakes, so the cyclotron tank suffered faults that opened under the enormous stress of the operating magnetic field. These bore the names of their discoverers—Van's vent, Henry's hole, Art's orifice. I won a measure of fame a few months later when I discovered a previously unknown fault; my colleagues christened it Luie's leak.

Just a few minutes of flaming the wax restored the vacuum that day. Jack's success wasn't a typical experience. I frequently spent an entire day searching for a leak, often demonstrating conclusively that no leak was possible; only a tank pressure ten times too high contradicted my careful eight hours of observation and hard work. Modern mass-spectrometer leak detectors had not yet been invented. Our primary leak sensor used natural gas sniffed out with a Pirani gauge, a thoroughly unsatisfactory arrangement that survived because no one knew any better way. Asphyxiation was added to the hazards of neutrons, gamma rays, high voltage, deuterons, and rats as we lay by the hour on the wooden oil-tank covers directing a stream of natural gas over the cyclotron's vacuum chamber.

Only shift operation made such a day tolerable. If I had been working alone, had spent an entire day looking for a leak, and faced another day of the same, I would have been desperate. But the operator on the next shift would bring with him the hubris of not yet having failed and wouldn't be at all disheartened by my eight-hour ordeal. Inevitably I arrived the next morning to read in the log that the leak had been found in a cranny I had gone over carefully four or five times. I wasn't any better or worse at leak detection than the next guy, and the shift system saved my sanity more than once.

I stood behind Jack for the rest of the afternoon, watching him bombard a target prepared by one of the laboratory staff. The cyclotron was then used almost exclusively to produce artificial radioactivity by deuteron bombardment. These primitive investigations involved measuring radioactive half-lives and the gross features of the emitted radiations and identifying the chemical element and atomic weight of the newly discovered isotopes. The chemical identifications used procedures straight out of elementary qualitative analysis.

The weekly schedule was posted on the control room wall every Monday evening so that we could choose our preferred shifts. Another schedule booked cyclotron time during which to make our targets radioactive. Married men got first call for the day shifts, but they also worked long hours in the evening, frequently accompanied by their wives. Each shift required a two-man crew—a crew chief and his assistant, or, as I now think of them, a pilot in command and a copilot. After several months of copiloting I enjoyed one of the proudest moments of my life: Franz Kurie casually suggested that I sign up as pilot in command.

Ernest Lawrence returned to Berkeley toward the end of May with the exciting news that he had raised nearly $70,000 to build a much larger cyclotron. He asked me what I had been doing since I arrived, and I told him that I had learned to operate and repair the cyclotron and was beginning to be able to tune the main oscillator. That was fine, he said, but he had a new job he wanted me to tackle immediately: designing the magnet for the new cyclotron. I pleaded ignorance of all things magnetic, to which Ernest characteristically replied, "Don't worry, you'll learn."

Learn I did. As a result of this brief conversation, I set up my workbench next to Ernie Lyman's and began building scale models of possible magnet designs and measuring their magnetic properties. Magnets today are designed on digital computers. In 1936 we had to use an analog computer—the model magnet operating at the same magnetic field as the full-sized machine. A model under those conditions will heat up according to the square of the scaling factor. Since my model was one-fifteenth the size of the proposed magnet, my uncooled copper coils heated up 225 times faster. I had to move with alacrity to catch a field reading between the time I turned the model on and the time its coil insulation would have burned out.

I had three principal variables and one fixed price—$50,000. The value I wanted to maximize was a product of the magnetic field and the radius of a particle's orbit; the maximum attainable particle energy is proportional to the square of this quantity. By increasing the cross section of the magnet yoke, I would increase the amount of iron required. But the heavier magnet would work more efficiently, reducing the requirement for copper. I concluded that I could achieve maximum particle energy when the cost of the iron I added equaled the cost of the copper I could eliminate. I turned countless four-inch pole pieces on the lathe trying to find the ideal tapered profile to give the highest value. (In all the prewar years I spent about 20 percent of my time in the machine shop. Modern young experimentalists learn little of machine-shop practice. Having spent a comparable amount of time at computer terminals, they know all about computers.) I finally decided that the yoke should be seventy-two inches wide and that the pole pieces should taper down to a sixty-inch diameter at the gap. Ernest went over my notebooks and announced that I had done a good job. The magnet was eventually built to my overall specifications. William Brobeck, who was responsible for the mechanical design, could equally well think of himself as the designer.

In the course of this design work, I came to understand better my shortcomings as a physicist. Before each measurement I pounded wooden wedges between the coils of my model to keep them from crashing together. Physics students know that a pair of coils with currents running in the same direction will be pulled together by well-understood magnetic forces. I eventually noticed that the wedges sometimes fell out during measurements. I was too busy at first throwing switches and watching galvanometer deflections to give this odd phenomenon any thought. I finally realized that my casual assumption was seriously defective—the magnet coils were in fact pushing each other apart. The iron cores caused the coils to separate.

I was upset that I had been too stupid to notice an obvious physical fact that had been staring me in the face for weeks. If I couldn't do a better job of observation, I wouldn't be much use to experimental physics. Not long before Wilhelm Roentgen discovered X-rays radiating from the fluorescing glass wall of a cathode-ray tube, the Oxford physicist Frederick Smith had learned from an assistant that photographic plates stored near such a tube had been fogged. Smith had simply told the assistant to move the plates. I had laughed at that story; now I realized I wasn't much of an observer yet myself.

I took stock then, comparing myself with my friends at the lab. By almost any standard, my training at Chicago had been atrocious. I had learned very little theoretical physics. My self-taught experimental skills were largely outmoded. In four years I hadn't had a suggestion from anyone—graduate student, postdoctoral fellow, or professor—to help me over the rough spots. I knew almost nothing of nuclear physics, radioactivity, radio-frequency engineering, high-gain amplifiers, modern vacuum practice, or electrical engineering.

From another point of view, though, my training had been extraordinarily good. I could build anything out of metal or glass, and I had the enormous self-confidence to be expected of a Robinson Crusoe who had spent three years on a desert island. I had browsed the library so thoroughly that I knew where to find the books I needed to learn almost anything I wanted to know. I've certainly learned more physics in the library than I ever did in course work.

It helped that people at Berkeley enjoyed instructing me. Someone passing through the cyclotron control room with a freshly bombarded target would often invite me along to watch him measure its radioactivity. I would fall in behind as he ran with his short-lived isotope to an electroscope room in the basement of LeConte Hall. From those forays I got a wonderful education in the practical aspects of radioactivity studies. Everyone had developed slightly different techniques; exposed to all of them, I was able to choose the most appropriate for my own later work. If I had come with training in nuclear physics, I probably wouldn't have received such a wonderful education. Since I was obviously both ignorant and eager, people felt secure enough to tell me everything they knew. Walking home each night to dinner, I would review what I had learned that day. Only after about a year and a half at Berkeley did what I taught others begin to outweigh what I learned.

I also brought from Chicago an intense curiosity about how everything worked and where everything was. I soon knew the contents of every drawer and cabinet in the Radiation Laboratory and had explored most of the rooms in LeConte. Among other things, I turned up the first small cyclotrons built in LeConte and several of Stan Livingston's early research notebooks. They're now on public display; if I hadn't found them in the back of a drawer on one of my systematic snooping operations, they would certainly have been thrown away.

Many of my colleagues who came to Berkeley on fellowships were so eager to publish a paper on radioactivity that they immediately claimed one or another element as their own and began bombarding it. Eventually they published a paper or two describing the new radioisotopes they discovered. But they learned very little about radioactivity in the process and became, in effect, technicians who could make measurements and do simple chemical analyses.

I chose another approach for two reasons, one altruistic and one selfish. I felt I should earn the right to use the cyclotron that others before me had developed at the price of enormous labor. As a result it was almost a year after I arrived before I signed up for a cyclotron bombardment. Moreover, I wasn't interested in adding yet another pedestrian radioactivity to the increasingly dull list. That was my selfish reason: I wanted to accomplish more interesting original work.

My standards were rising fast because of the reading program I was pursuing at night at home. Part of that program was a systematic review of the literature. The only way that I could really learn nuclear physics, I concluded, was to read *everything* that had been written on the subject. I arbitrarily decided that the beginning point of my reading should be Ernest Rutherford's 1919 disintegration of the nitrogen atom by alpha-particle bombardment, an event that changed nuclear physics from an observational science like astronomy to an interactive one. I carried home one annual volume after another of the old *Philosophical Magazine, Nature,* the *Proceedings of the Royal Society,* the *Proceedings of the Physical Society of London,* and, for the years after nuclear-physics research caught on in the United States, the *Physical Review.* I supplemented my reading of the literature in English by examining summaries in *Science Abstracts* of French and German work. If I saw something interesting there, I looked up the full-length report.

Every night for more than a year I scoured the bound volumes. It was exciting to read the basic papers to which I had seen references in books. I especially remember reviewing Rutherford's 1920 Bakerian Lecture, in which he first proposed the possibility of a neutron—an elementary particle not discovered for another twelve years, until which time the only known constituents of matter were electrons and protons.

Another activity that helped change me from a raw postdoctoral assistant into a professional nuclear physicist was Ernest Lawrence's journal club. Every Monday night at 7:30 we turned the cyclotron off and met in the old LeConte library. Since most physicists find they don't really understand a subject until they've discussed it with their colleagues, talking physics is essential to doing physics. So we gathered weekly to discuss the nuclear-physics literature. By 1936 we more often discussed new discoveries at the Radiation Laboratory than reports of work at other laboratories. Lawrence made a point of not announcing the evening's program in advance. The speaker could be a graduate student giving his first public report or Robert Oppenheimer, the distinguished American theoretical physicist, announcing his theoretical discovery of neutron stars. It was considered a serious breach of etiquette to ask who the evening's speaker would be. Many years later, when that rule was relaxed, attendance fell away and the journal club closed its doors. (A modern version of Ernest's journal club has been held in my home almost every Monday night for the past twenty-seven years. Many of my students have followed my lead, so there are now "Luie meetings" all over the world.)

In its heyday the journal club provided the principal place where experimentalists and theoreticians interacted. The Radiation Laboratory was completely focused on experiment. It offered only one desk where paperwork was possible. Robert Oppenheimer and his remarkable group of graduate students kept the experimentalists up to date. They made us all feel at home with the theorists' jargon even though we often didn't understand in detail what they were talking about. They were keenly interested in what went on in the laboratory, and whenever an experimentalist needed help with theory they were ready to pitch in.

Among Robert's students during this period were Bob Serber, Bob Christy, Julie Schwinger, Willis Lamb, George Volkoff, Hartland Snyder, Phil Morrison, Sid Dancoff, and Leonard Schiff. This was probably the finest group of young theorists in the world. As a result I had first-rate theoretical advice available in the time it took to walk from the laboratory to Robert's bullpen in LeConte. Each theoretician had his own private cubbyhole, but the group usually crowded with Robert into one of the rooms at LeConte to watch and comment as someone worked equations on a blackboard. Robert's permanently floating crap game was in every sense the theoretical counterpart of the Radiation Laboratory; the theorists similarly worked together, helping each other enthusiastically and happily with criticism and advice. Every spring the Oppenheimer crews would load into their cars and float down to Caltech, where Robert would teach a formal course or two and continue leading around-the-clock discussions at the blackboard.

The other important component to my self-help program was a detailed study of three articles that appeared in the *Reviews of Modern Physics* in 1936 and 1937. These, by the German-born theoretician Hans Bethe and his colleagues Bob Bacher and Stan Livingston (by then moved from Berkeley to Cornell, where Bethe also taught), made up what came to be famous as Bethe's Bible, a thorough and original reexamination of the entire corpus of nuclear physics as it was then understood. When men of wealth want to learn about a new field, they sometimes commission an expert to write a monograph on the subject especially for them. I felt that Bethe had performed just such a personal service for me; the articles seemed to have been tailored to match precisely my ability to absorb the knowledge they contained. I saw the first article in the physics library shortly after I arrived in Berkeley and sent a check to the American Physical Society for my own, personal copy of that issue of the journal. The three articles together ran to 468 pages.

Many of the experiments I performed in those days were provoked by Bethe's conclusions. If he said a phenomenon would never be observable, I wanted to prove him wrong, which would make both of us happy. In several significant instances over the next four years, I did.

THREE

Coming into My Own

MY TRANSITION from apprentice to professional was not without incident. I fitted the Radiation Laboratory's style comfortably, and the staff regarded me as hardworking and useful, but like my fellows I was sometimes consigned with good reason to the laboratory doghouse.

We usually disgraced ourselves by cutting corners, trying to work faster than prudent operation warranted. Franz Kurie, for example, once put the cyclotron out of business for several days by forgetting to remove from the vacuum chamber the hardwood blocks we used to support the dees when we reassembled the cyclotron after repairs. The vacuum pumps began evacuating the reassembled tank, which allowed the blocks to outgas, and we noticed that the pumps were laboring. We suspected a leak, of course. We wasted most of a day looking for it. Eventually the wood outgassed sufficiently to maintain a decent vacuum. But when we applied the radio-frequency voltage to the dees it heated the blocks, they dumped new gas into the vacuum, and the pumps began laboring again. This unusual cycle repeated itself. Finally we had to remove the chamber. Then we found the blocks, like sponges forgotten in the belly of a surgical patient. Among the modest gifts exchanged at the laboratory Christmas party that year, Franz collected enough wood to warm his living room through a harsher winter than Berkeley's.

I mistakenly buggered an interlock. (Those familiar only with the vulgar meaning of this verb should know that *Webster's* defines "bugger up" as "put into disorder.") Interlocks are devices that protect machines and people from accidents. An interlock disconnects the electrical power from a television set, for example, when the back of the cabinet is removed, to guard the unsuspecting from contact with the 30,000-volt circuit that powers the color tube. A television repairman has to bugger the interlock and stay away from the high voltage to fix the set. Buggering interlocks is accepted practice, then, but only if you are certain you know what you're doing and only if you unbugger the interlock when you're through.

On cyclotron crew one day I went through the routine of opening a valve that admitted cooling oil into the upper coil tank. A turning of the valve handle simultaneously lifted a weight attached to a length of clothesline. An electrical relay normally kept the weight from falling when the valve was fully open. I had never examined the plumbing for this part of the cyclotron, because it was one of the few systems that hadn't broken down in my many months of crew duty.

The relay refused to work. The weight dropped and closed the valve. Since I couldn't operate the cyclotron with the valve closed, I buggered the interlock by putting a clamp on the clothesline just above the pulley. Then I turned on the cyclotron and ran it until my shift ended. After dinner I returned to the laboratory with Gerry to find an extremely angry evening crew chief. In front of my bride he thoroughly chewed me out, after which he led me to the scene of the crime in the basement below the cyclotron, where I had often spent hours working in the two-foot crawl space between the dirt subfloor and the wooden floor beams. We found the subfloor covered now with transformer oil; the oil valve had been interlocked to prevent oil from leaking from the top tank and overflowing the bottom tank. Positive oil pressure normally kept the valve open; the oil pump had apparently stopped working. Because of my tampering all the oil in the top tank had drained out, exposing the upper coils, and it was only good fortune that spared a burnout. If the coils had burned out, the magnet would have been shut down for repairs for weeks. I was saved from such absolute disgrace because the cyclotron had been operating only marginally at the time.

That was a close call; it taught me to be much more careful in making changes on collective equipment. I had been a lone wolf in Chicago. If I damaged my apparatus there, I hurt only myself. Now my calculated risks could jeopardize my friends as well. I learned my lesson and thereafter buggered no interlocks unless I understood their function.

Group operation under the conditions of that era exposed us to other hazards as well. My first published work at Berkeley concerned removing the ion beam of the cyclotron from the magnetic field. It was possible to direct the beam of deuterons as it came out of the dees through a window of thin platinum foil into the open air, where it made a luminescent blue glow several inches long. Ernest Lawrence's love affair with the deuteron beam was legendary. He led every laboratory visitor into the cyclotron room to see the luminescent beam. As the beam current increased with improved equipment and larger cyclotrons, the blue glow became visible even in a fully lighted room.

But we seldom kept adequate records of cyclotron modifications. Some months after I arrived in Berkeley, the chamber of the 27-inch cyclotron was removed for major improvements. Rather than idle the instrument, we temporarily reinstalled the previous chamber. It didn't permit the beam to leave the vacuum chamber, but it had a reentrant viewing cylinder that could be operated either evacuated or filled with air. With air the glowing ion beam was visible; without air there was nothing to see, but the beam impacting invisibly on its beryllium target made a flood of neutrons.

Alvin Weinberg, a friend from graduate-student days in Chicago, dropped in unexpectedly one afternoon to see the laboratory. It fascinated him as much as it had fascinated me when I first toured it. Before he dropped in, I had watched the beam streaking blue across the cylinder; now I invited Al to watch with me. I asked the crew chief to turn on the cyclotron, and then Al and I knelt leaning on the lower coil tank with our heads about a foot from the viewing window. After a time I shouted to the crew chief that he didn't have a beam. He shouted back that he did. I said it wasn't enough to see, and the crew chief said it ought to be more than enough. Out of the corner of my eye I then noticed the beryllium target glowing bright red. I grabbed Al and pulled him away from the cyclotron and at the same time shouted to the operator to turn off the machine. In the hour since I had watched the beam alone, someone had evacuated the air from the viewing cylinder; what I saw from a distance of one foot was the beryllium target heated by a strong current of 5 MeV deuterons. Al Weinberg and I probably received the largest dose of fast neutrons anyone had ever experienced up to that time. Luckily we escaped without injury and both went on to full scientific careers.

One of the experiments I carried through by questioning the infallibility of Bethe's Bible shows the extent to which I was learning my trade. Among the earliest discoveries of turn-of-the-century research into radioactivity was that of the emission of beta radiation, which was demonstrated to be composed of energetic electrons. With the emission of an electron, a beta emitter transmuted itself atom by atom into an isotope of the same mass number but with an atomic number one unit higher.

What happened within the nucleus itself during beta decay was obscure until the discovery of the neutron in 1932 and Enrico Fermi's classic 1934 theory of beta decay, one of the landmarks of nuclear physics. Before the neutron was discovered, theorists had assumed that the electrons that came out in beta decay existed beforehand in the nucleus along with positively charged protons. (If you see a man walk out of a house, you naturally assume he has been inside the house a moment earlier.) But there clearly wasn't room to pack many electrons into the diminutive nucleus, so the theory had been in trouble. Fermi straightened things out. He modeled his theory after the processes of electromagnetic radiation. Just as a photon of light is created by an atom at the instant it is emitted, he assumed that the electron emitted by a nucleus during beta decay is created at the moment of decay. To satisfy certain other requirements of the reaction, Fermi also postulated the simultaneous creation of a new particle, previously proposed on theoretical grounds, which he named the neutrino. Massless and electrically neutral, the neutrino was an elusive customer, and in fact vast numbers of neutrinos ejected from the sun stream constantly through the earth and all its inhabitants. Demonstrating its existence experimentally was one of the more monumental efforts of physics, not accomplished until long after the Second World War.

Fermi's theory predicted beta decay not only by the emission of negative electrons but also by the emission of positive electrons—positrons, discovered as a component of cosmic rays by Carl Anderson at Caltech in 1932. Positrons also were created at the moment of decay and could be thought of as particles or, equivalently, as holes in the continuum of negative-energy electrons. Under certain circumstances it should be possible for an excited nucleus to decay by capturing one of its own orbiting electrons. No positron would be emitted, but a transmutation would nevertheless occur.

The problem Hans Bethe addressed was how to demonstrate such electron capture by nuclei—K-capture, it's called, because the nucleus most easily captures electrons from its innermost electron shell, the K shell. Among particles, Bethe pointed out, only the unobservable neutrino was emitted in K-capture. He thought one might detect the process by counting positrons over a long period and comparing their number with the number of atoms formed in a measured sample of material by decay. If the two numbers differed, the difference could be attributed to K-capture. That would be a difficult and time-consuming process, though one of my students eventually accomplished it for sodium 22 turning into neon 22.

Bethe had been a distinguished theorist of atomic structure before his interest shifted to the nucleus, but his 1936 *Review* paper failed to comment on the fact that a transmuted nucleus immediately after K-capture finds itself inside an atom with one of its innermost electrons missing. An electron from an outer shell would immediately fall into such a vacancy; in doing so, it would emit a characteristic K X-ray. A neutrino might be elusive, but X-rays we knew how to detect.

An experimenter at Niels Bohr's Institute for Theoretical Physics in Copenhagen saw that possibility first. He searched for X-rays in a cloud chamber but had the bad luck to look for electron capture in a radioisotope that didn't decay by that means. As soon as I read his paper, I realized that his was the proper way to look for K-capture and immediately started designing an apparatus structured to find soft K X-rays mixed in among the positrons and gamma rays that would accompany electron capture in vanadium 48.

By putting the radioactive sample in a moderate magnetic field—I used the model magnets I had wound for my 60-inch-cyclotron studies—I could curl the beta electrons out of the way before they reached my detector. The K X-rays I was looking for were easily absorbed within a few centimeters of argon; gamma rays from positron emission, on the other hand, would pass through the argon with very little loss of energy. So I set a trap to catch soft X-rays but not hard gamma rays. I designed and built a Geiger counter filled with argon at normal atmospheric pressure, but instead of blowing the usual glass envelope I enclosed the argon in a cellophane cylinder with walls only a thousandth of an inch thick. I designed my X-ray trap for the very small range of K X-ray energies produced by chemical elements in the neighborhood of titanium, down to which vanadium 48 decayed. To prevent such low-energy X-rays from being absorbed by the air between the titanium sample and the Geiger counter, I enclosed the equipment in a helium-filled bag. Finally, I devised a wooden holder with which I could slide frames of extremely thin aluminum foil in and out of the X-ray beam. By measuring the absorption of the X-rays in aluminum, I could see if they matched the standard measurements of titanium K X-ray absorption in aluminum.

The experiment involved taking two thousand counts at each of seven values of absorber thickness and repeating each setting eight times. I found what I was looking for, or thought I did, and proceeded then to study the process in iron, nickel, copper, and zinc. I was now the sole occupant of a pleasant research room in the basement of LeConte Hall. Since radioactivity research involves hours and hours of waiting for a sample to decay, I had time to become acquainted with a graduate student who shared an adjoining room, C. S. Wu—Gigi, to her friends—one of the most talented and beautiful experimental physicists I have ever known (she was elected president of the American Physical Society six years after my stint in that post).

I found X-ray emitters in all these samples, so it seemed that K-capture was common in positron decay. Then I encountered another process in another isotope that produced characteristic K X-rays but didn't require electron capture. To prove the existence of a new phenomenon, all alternative explanations have to be excluded. I was in a pickle. I was fortunate to have recognized the problem myself before I published my claim and someone else shot me down. Because of my experience with X-rays at the University of Chicago, I was able to resolve the ambiguity by conclusively demonstrating K-shell electron capture in gallium. The report I published in 1938 was definitive, and the capture of orbital electrons by nuclei continues to be associated with my name.

Since the discovery of artificial radioactivity, in France in 1934, the cyclotron had been used primarily to induce radioactivity in most of the known chemical elements. The Radiation Laboratory used the cyclotron as a radioactivity factory first of all because great numbers of new radioisotopes could be discovered that way with very little effort. The gold rush of isotopes lasted four years, and only after the Second World War was there time and adequate instrumentation for the researchers to follow through with detailed analyses of all the radiations the hundreds of radioisotopes emitted.

Contingency also affected the choice; the cyclotron was available only a small fraction of the time. We spent many hours finding leaks, adjusting equipment, repairing oscillators, and developing cyclotron technology. Lawrence's job as director also required him to consider the necessities of public relations. He wanted radioisotopes used in all branches of science, and this led him to missionary work with the other Berkeley science departments. We recognized the importance of these activities, but after spending days in cyclotron repairs we grumbled when a physiologist or a biologist turned up to claim the fruits of the first bombardment. We grumbled among ourselves, that is; we knew the strength of Ernest's convictions and were much too loyal to allow outsiders to discover our ambivalence.

Once Lawrence came close to violating our constitutional guarantees against cruel and unusual punishment. For most of his long, fruitful career Ernest Rutherford, Lord Rutherford of Nelson, had expressed his preference for experiments managed on modest equipment assembled with legendary Cavendish string and sealing wax. He had quarreled about the matter with James Chadwick, the discoverer of the neutron and a Rutherford protégé, when Chadwick had wanted to build a cyclotron—quarreled unhappily enough that, after sixteen years at the Cavendish, Chadwick had gone off to the University of Liverpool and there built England's first cyclotron. Rutherford had now finally seen the light and had sent John Cockcroft to Berkeley to learn about cyclotrons. The money became available when Peter Kapitza, a Russian physicist working at the Cavendish, had been detained in his homeland by the Soviet government while attending a conference and forbidden to return to England. Kapitza had raised funds in England and despite Rutherford's prejudices, which he alone seems to have been able to circumvent, had built and equipped in the Cavendish courtyard a massively instrumented new laboratory devoted to research into powerful magnetic fields. Rutherford appealed to the Soviets to allow Kapitza to return to England and continue his scientific work, to which they responded that they would be equally pleased to have Rutherford in Russia. They agreed, however, to buy all of Kapitza's apparatus. The value of the equipment came to some £30,000. Part of that fund was earmarked for a cyclotron.

Few occasions can have given Lawrence more pleasure than his receiving an emissary from the great Ernest Rutherford himself. Lawrence had known John Cockcroft since Cockcroft's visit to Berkeley in 1933. They had found they had much in common and had immediately become friends. Ernest personally showed John the laboratory, taught him how to operate the cyclotron, and told him what he could expect to see when the cyclotron tank was pulled for the inevitable repairs.

As Cockcroft's departure approached, however, the cyclotron for the first time in memory stubbornly refused to break down. Since Ernest had promised John a look inside the chamber, I was prepared to be ordered to shut down and pull the vacuum tank for inspection. Instead Ernest concocted a truly diabolical subterfuge. He turned off the compressed air that cooled the glass insulators that supported the dee stems and their cantilevered dees. While Ernest operated the cyclotron and John looked over his shoulder, I watched with mounting horror as the radio-frequency fields heated the glass insulators to the softening point. One of them collapsed. Air rushed into the vacuum chamber. We all dashed to the rescue, just as we would have done had the "accident" been accidental. Ernest pointed out the site of the failure to his guest and blithely informed us that they would return after dinner to watch the reassembly. We were speechless.

Bethe's Bible and a long discussion with my father during one of his visits to Berkeley led me to another experimental interest. Dad and I talked about the identifying of problems that are really worth working on. Looking back on his career, he said, he realized he should have taken time out more often to sit down and think about what he was doing. He had always felt pressed to follow a mental list and had missed important leads. He had once thought hard about pernicious anemia, he told me, in those days a fatal disease. He had fed ground liver to a few anemic dogs with no certain results and then moved on to other pressing projects. He was the more unhappy a year later when his immediate superior at the Hooper Foundation, Dr. George Whipple, demonstrated that massive doses of raw liver restored the health of anemic dogs. Whipple's work established a standard treatment for human pernicious anemia that won him and two colleagues the 1934 Nobel Prize in medicine. Dad never claimed he might have won that Nobel Prize, but he did say more than once that he would have been a better researcher if he had occasionally let his mind wander over the full range of his work. He advised me to sit every few months in my reading chair for an entire evening, close my eyes, and try to think of new problems to solve. I took his advice very seriously and have been glad ever since that I did.

By the spring of 1937 my literature survey was far enough along to allow me to put my father's suggestion into practice. Two gallbladder attacks offered unwanted assistance. Doctors think of gallbladder patients as female, fair, fat, and forty. I was male, fair, thin, and young, an uncomfortable exception to the rule. My second attack caused me intense pain, which massive doses of morphine barely attenuated. In bed for some time recovering from jaundice, I had the leisure to follow my father's advice.

Niels Bohr had delivered the prestigious Hitchcock Lectures at Berkeley that spring. The audience at the first lecture overflowed the auditorium but quickly dwindled to a hard core of physicists and chemists thereafter. Bohr whispered his lectures, *dropping* his voice to emphasize a point. Wandering from the podium microphone to the blackboard, he was barely audible to those in the first few rows. I was fortunate to find a seat in those rows for all five lectures. His work on the liquid-drop model of the nucleus impressed me deeply.

It led me to think about the problem of slow-neutron capture, which happens with greater frequency at certain neutron energies than at others, a phenomenon known as resonance. Bohr drew extensively on our meager experimental knowledge of neutron capture resonances. All we could establish at that time was the rank order in resonant energy between one element and another. If optical spectroscopy were as primitive as neutron spectroscopy, Ed McMillan had quipped in a lecture the preceding year, instead of using prisms and diffraction gratings we'd be using colored shards of broken beer bottles.

I meditated at length on the problem and decided to look into the possibility of measuring neutron energies by measuring their velocities. As far as I know, I was the first person to think of the method, now commonplace, of measuring time of flight. The problem with neutrons was sorting out fast neutrons from slow. Under most circumstances they arrive at a detector jumbled together. Cadmium is a powerful absorber of very slow ("thermal") neutrons; by taking a series of measurements of a neutron beam with and without cadmium sheets interposed, investigators had been able to identify the effects of slow neutrons as the difference between the two conditions. But slow-neutron absorption experiments had been compromised by the ever-present and apparently unavoidable fast-neutron background.

Soon after the discovery of slow neutrons by Fermi's group in Rome in 1934, John Dunning, Emilio Segrè, and their colleagues at Columbia showed, by using rotating shutters, that thermal neutrons really moved as slowly as hydrogen atoms at room temperature did. But they couldn't get rid of their fast-neutron background, so their important experiment didn't lead anywhere. But the fact that slow neutrons take an appreciable time to travel distances on the order of meters —1/240 of a second for eight meters, in the case of a thermal neutron —was the basis for the method I devised.

I had hoped to explore the "resonance region," where neutrons with energies of a few electron volts are preferentially captured. After many attempts to turn the cyclotron beam off fast enough to work in this region, I had to give up; the ion source we were using didn't permit it. So I backed off and used the slower "modulation times" that let me work with what was the first beam of pure thermal neutrons—in fact, thermal neutrons with a wide range of effective temperatures, down to very low values.

I arranged to turn the cyclotron beam on and off 120 times per second. That generated fast neutrons in the beryllium target in pulses about half the time. Near the target I placed a block of paraffin wax. Some of the fast neutrons passing through the paraffin would collide with its hydrogen atoms and slow down. The paraffin block faced one end of a long cadmium tube; at the other end, some eight meters away, I fitted a boron trifluoride ionization chamber that could register the arriving neutrons and display them through a linear amplifier on an oscilloscope screen. The linear amplifier I arranged to be sensitive only when the cyclotron beam was off.

Since fast neutrons have a very short life span—a mean of less than 10^{-5} second—I could turn on the cyclotron beam for a few milliseconds sixty times a second and detect only room-temperature neutrons. The faster neutrons would have passed the detector before it was sensitive, and the slower ones would reach it after it had been deactivated. Operationally speaking, I now had a beam of pure thermal neutrons. People who had worked with neutrons were amazed to see that when I interposed a thin sheet of cadmium between the cyclotron target and my neutron detector, the counting rate dropped to zero, as if the cadmium were a sheet of black paper intercepting ordinary light and as if fast neutrons didn't exist.

In 1937 no one thought it was possible to make a beam of pure thermal neutrons. The only materials anyone used to slow down neutrons were those containing hydrogen. The materials unfortunately also absorbed thermal neutrons more effectively than they did fast neutrons. So in hydrogenous materials fast neutrons were always present among the slow. Later, when Fermi and Leo Szilard in the United States and Frédéric Joliot and his colleagues in France were trying independently to invent a nuclear reactor, they realized that other materials that absorbed fewer slow neutrons would serve their purposes better than water or paraffin. Necessity was the mother of invention; the U.S. tried beryllium and then carbon and found that neutrons could be slowed to thermal velocities at which they could travel long distances without appreciable loss. It only took a lot more collisions to do the job. The French physicists hoped to use heavy water, but before they could accumulate enough of it the Second World War intervened. Szilard and Fermi concluded that graphite was best suited for the job, and the rest is history.

My "pure" thermal beam achieved its effects differently from the graphite columns Fermi later devised to produce true beams of pure thermal neutrons, but operationally it was the first such machine. It had strange manifestations. If I arranged to delay my measurements by ten milliseconds instead of five milliseconds, a guest observer would be startled to find that the apparent temperature of my neutron beam had dropped from room temperature, 300 degrees Kelvin, to 75 degrees Kelvin, which is below the temperature of liquid air. These cold neutrons took twice as long to reach the detector. With half the velocity of their predecessors, they were only one-fourth as energetic and therefore that much cooler.

To pulse the flow of neutrons, I had to turn the cyclotron oscillator on and off. And to make sure I was really stopping and starting the deuteron beam, I made a test in which I ran the deuterons into a thin ionization chamber that just happened to have lithium fluoride smeared on one of its electrodes. The beam current dropped to zero about two milliseconds after the oscillator voltage peaked, as expected, but a background of heavily ionizing particles appeared uniformly across the oscilloscope screen. These particles were obviously related to the deuteron beam because when the cyclotron was turned off, the background decreased with about a one-second half-life.

It took me a few minutes to realize that I had independently rediscovered artificial radioactivity, and by a method that should have led to its discovery at Berkeley before Frédéric Joliot and Irène Joliot-Curie discovered it in Paris in 1934. (That discovery had been denied to the Berkeley group because they thought they couldn't make good Geiger counters—because, as we now know, their whole laboratory was artificially radioactive. But they were skilled in the use of thin ionization chambers of the kind that I was now using and that were insensitive to beta rays but very sensitive to the alpha particles I was now seeing. And lithium is almost unique in giving delayed alpha particles.) Bombarding the lithium in my detector made it radioactive; it responded by emitting alpha particles following beta decay. I had just made a really extraordinary discovery. If I hadn't been three years too late, I would certainly have told someone about it.

I did look into why Berkeley missed it. To make accelerated deuteron beams, G. N. Lewis, Ernest Lawrence, Stan Livingston, and Malcolm Henderson had used equipment almost identical to the equipment I was now using; they had described this work at the meeting of the American Physical Society I had attended during the Chicago exposition in 1933. Lewis, the dean of American chemists, had concentrated the heavy water from which the deuterium was electrolyzed, so his name was first on the paper even though he wasn't very familiar with the cyclotron or the detection equipment. Lawrence and Livingston were undoubtedly busy keeping the cyclotron running. So the Berkeley person who missed discovering artificial radioactivity, for which the Joliot-Curies won the 1935 Nobel Prize in chemistry, was Malcolm Henderson.

I couldn't understand why Malcolm missed seeing what I had just observed. Like me, he had used a thin ionization chamber with deuterons hitting a lithium target. Like mine, his cyclotron switched on and off sixty times a second (to save money in those days, Berkeley used "raw AC" to power its oscillator). I went back over Malcolm's earlier papers to familiarize myself with his techniques. If he had displayed his detector signal on an oscilloscope, he would have seen the effect, just as I had. I doubt that he could have tuned up his apparatus without an oscilloscope. He may have missed so important a discovery because he simply didn't expect it and therefore didn't watch his oscilloscope as closely as I had. Or he may have turned it off to increase its useful life; oscilloscopes were expensive in those days.

I had now made two major discoveries in nuclear physics: K-electron capture and artificial radioactivity. I haven't ever mentioned the second of the two before; I do so here only to show that I had now learned my trade and noticed things I hadn't expected to see, which is, of course, essential to scientific discovery.

Enrico Fermi lectured on slow-neutron diffusion theory at Stanford University, across San Francisco Bay in Palo Alto, in August 1937. His brilliance as a lecturer was astonishing. I had never heard such lucid lectures, delivered without notes but as polished as if Fermi had been reading them from a book. My friends and I attributed his flawless performances to careful rehearsal. Later, when I worked with him, I understood that he simply knew his subject so well that he could improvise a finished presentation only minutes before he walked into the room.

Fermi visited the Radiation Laboratory, and Lawrence introduced us. Later we spent a pleasant Sunday on Ernest's cabin cruiser showing Enrico the bay. Ernest had bought his cruiser from a ferryboat captain who had bought it in turn from my dentist uncle Harold, who liked to work with his hands on a large as well as a small scale and had built it himself. Ernest had won the Comstock Prize that year. It is awarded every five years by the National Academy of Sciences and was the first of his many medals and awards. It led to a *Time* magazine cover story. He used the money to buy the boat.

In 1937 I received interesting job offers from Bell Laboratories, Cornell, Chicago, and the University of Minnesota, but so long as I was wanted at the Radiation Laboratory I intended to stay. In November, Ernest told me I would be appointed a Berkeley instructor beginning the following summer. That was good enough for me.

We devoted 1938 to building the Crocker Laboratory and its big, 60-inch cyclotron. One memorable February day the biologists and their rats moved into Crocker and at last freed the old wooden laboratory of their stench. We still spent weeks making phosphorus 32 for John Lawrence, taking time away from physics research. John had encouraging initial success with several leukemia patients who ingested orange juice spiked with radiophosphorus. Later these early cures turned out to be the statistical fluctuations that make medical research so difficult. John then demonstrated that polycythemia vera, another fatal blood disease, could be cured by the drinking of a radiophosphorus cocktail, still the standard treatment. In May, Ernest inaugurated an owl shift at the cyclotron, from 11 P.M. to 3 A.M., to make the phosphorus with a fourteen-day half-life. In July the laboratory went to around-the-clock operation to satisfy biomedical needs.

John Lawrence and Joe Hamilton were the two medical doctors who spent most of their time in the laboratory. I certainly owe my life to John; had he not come to Berkeley in 1935, I would probably have died of radiation sickness long ago. John's first experiment was designed to determine the degree of radiation hazard associated with neutron bombardment; the neutron fluxes at the Berkeley cyclotron were enormously more intense than those available at any other physics laboratory in the world. John had a brass cylinder made just large enough to hold a mouse. Rubber tubes carried a stream of air to the mouse when the cylinder was placed close to the beryllium target in the 27-inch cyclotron. After a fifteen-minute exposure to the target neutrons, the mouse was removed. It was dead. When I arrived on the scene several months later, people were still talking about the shocked silence that greeted that observation. By then John's secret had leaked out: the mouse died of asphyxiation; someone had forgotten to turn on the compressed air. John subsequently showed that fast neutrons were several times more dangerous per unit of radiation dose than the gamma rays from radium. As a result, the cyclotron control table had been moved as far from the target as possible. As Ernest continued his relentless drive to raise the beam current in the cyclotron, John decided that the cyclotron operators were getting too high a dose of neutrons. He led a campaign then to surround the entire machine with water tanks. I had been a cyclotron operator for several months before the tanks were put in place, so my debt to John Lawrence is large, and frequently remembered.

Joe Hamilton became more deeply involved in the techniques of nuclear physics than John, but they both did important experiments in medical physics. In 1937 Joe asked me if I would like to collaborate on an experiment he had been thinking about to use radioactive sodium, which has a half-life of fifteen hours, to measure the absorption of sodium ions from the stomach wall into the bloodstream. He proposed that each of us would drink a millicurie of sodium 24 while holding a Geiger counter shielded inside a lead-lined box and watch the counting rate increase with time. I told him that I would calculate the dose that night; if it looked all right to me, he could count me in. Radiation doses increase as the inverse square of the distance, so they shoot up fast as the radioactive source approaches the body. But what happens if the source actually enters the body? That's the kind of problem Newton invented the calculus to solve; I was soon able to conclude that Joe was proposing a reasonable experiment (it wouldn't be permitted today, of course). As medical tradition dictates, Joe did the experiment first on himself. A few hours later I became the second guinea pig. As far as I know, it was the first experiment of its kind.

When Ernest gave his Faculty Research Lecture a few months after Joe's experiment, he demonstrated it onstage—using a tenth of a millicurie of sodium 24 and with Robert Oppenheimer as the guinea pig.

In May 1986, after the Soviet reactor fire at Chernobyl, U.S. newspapers carried daily front-page stories about the radioactivity deposited on our soil by its fallout. The typical amount of radioactivity measured was ten picocuries per liter of rainwater. The one millicurie Joe and I each considered safe to drink was a hundred million times stronger than the ten picocuries that apparently scared the American public in 1986 sufficiently to cause many citizens to cancel long-held European travel reservations. I have been outraged for years by the media's exaggeration of the dangers of trivial amounts of radioactivity. But the public reaction may be the good news about Chernobyl. Both our leaders and those of the Soviet Union saw then how their citizens react to the presence of even tiny amounts of radioactivity. I can't believe that either leadership would ever attempt a preemptive nuclear strike, which would certainly be answered in kind, when the resulting levels of radioactivity would be measured in kilocuries or megacuries, a million or a billion times more dangerous than the one millicurie Joe and I each drank, which was in turn a hundred million times more than the amount that frightened Americans in 1986.

It's good news as well, I think, that some of those who guarded the Chernobyl reactor left their posts because they feared exposure to radiation. We have evacuation plans in this country to deal with reactor accidents. I have long felt that these plans could never be implemented, because the police, who are expected to stand their posts and direct traffic to avoid gridlock, would most probably depart immediately for home to deliver their spouses and children to safety. To condition soldiers to remain in hazardous situations takes long training under wartime conditions; I'm convinced that, lacking such wartime training, neither Soviet nor American officers would stand their posts. Both populations of civilians would panic. Chernobyl offers evidence, though it was a minor radiological incident compared to what a nuclear exchange between the two superpowers would be. So I find it difficult to believe that the leaders of either country could push the button after seeing what happened at Chernobyl.

Radioactive waste burial is a related issue. No U.S. governor will allow his state to become the dumping ground for other states; as a result, the waste keeps piling up in tanks and reactor pools. The solution I favor has been tested experimentally by British engineers: burying waste some one hundred feet below the sea bottom. It couldn't possibly do any harm there. I speak as a recognized expert on the migration of materials in seafloor limestones; I've spent much of the past five years examining the record of such migrations over tens of millions of years of geological time and I would qualify to give expert testimony on the subject in any court of law.

The usual extent of the migration is less than one foot, and we have never seen material move as much as five feet. The British engineers showed that gravity-powered torpedoes (steel pipes one foot in diameter and ten feet long with pointed noses and tail fins) reach terminal vertical speeds large enough to drive them one hundred feet into the sea floor. One can be quite sure that none of the radioactivity so buried would ever find its way back up into the ocean, but even if it did, it would be lost in the approximately one billion tons of uranium already in the ocean.

I hope that someone who is worried about waste disposal and who is in a position to do something about it will see to it that this simple and elegant solution is implemented. There are probably treaties to prohibit such a disposal method. They must have been written by people who were unaware of the stability of seafloor sediments. I would advise the governors to stop fighting the federal government and use their energies instead to encourage this approach.

For me this was a period of intense concentration on physics and on laboratory operation. I had progressed from crew member to shift captain and to responsibility for weekly operations, which I suppose corresponded to admiral. I seldom read a newspaper; I followed the world by reading *Time* on weekends. My teaching started in August, and I rediscovered the truism that you don't begin to understand physics well until you teach it. I made any number of mistakes. I admitted them freely. My students were tolerant, the more so since they were taking the undergraduate courses I taught only to qualify for medical school. I was fortunate to make my mistakes and learn from them before such an undemanding group.

I began a long program of experimental work that year with Felix Bloch, the Stanford theoretician. He was the first person I had worked with of cosmopolitan background, and I learned more from him than theory. I listened in amazement to his conversations with Emilio Segrè, who had recently left Italy ahead of the new anti-Semitic laws Benito Mussolini imposed to placate Adolf Hitler. The two men would start in English but then switch suddenly to German, French, or Italian. When Gerry visited her family in the summer, I often dined with Felix at the faculty club. He relaxed before dinner at the concert grand playing classical music from memory. I relaxed before dinner by playing three-cushion billiards with Don Cooksey; we were two of the best players in the club.

When Ed Purcell introduced me before my retirement talk as president of the American Physical Society in 1969, he said he thought the experiment that Felix and I pursued in first measuring the magnetic moment of the neutron was one of the most difficult ever done. (It certainly wasn't easy.) Ed was following the tradition that the incoming president of the Physical Society introduce the retiring president. A year earlier I had that pleasure when I introduced Charlie Townes, the inventor of the laser. I had shown Charlie a photograph the day before that I thought would make a memorable conclusion for his lecture. I had just flown in from a NASA meeting at the Jet Propulsion Laboratory, where I had been given a day-old photograph taken by the first camera located on the moon. Observatories in Texas and Arizona had been waiting for the camera to land and knew its lunar coordinates. They immediately aimed laser beams at it. The photograph I showed Charlie was of the crescent earth, with two bright points of light on the dark hemisphere. Charlie agreed it would make a "socko ending" for his talk, and I volunteered to have a slide prepared. At the appointed time Charlie showed the slide. The two spots were nowhere to be seen. Only after that downbeat ending did we realize what had happened. The commercial photographer who made the slide had assumed the photograph must be the moon. If so, then the white spots must be blemishes, which a little black ink could repair. No doubt his English teacher had taught him, as mine had taught me, that Samuel Taylor Coleridge made a similar error in *The Ancient Mariner* when he referred to "The horned Moon, with one bright star/Within the nether tip. . . ."

In the midst of our long and difficult experiment, I managed another experiment that was quick and very important. Until 1932 hydrogen was believed to consist of a single isotope of atomic mass 1. Helium was believed to consist of a single isotope of atomic mass 4. Then Harold Urey at Columbia University found a rare isotope of hydrogen with mass 2, now known as deuterium. A year later Ernest Rutherford and his coworkers at the Cavendish discovered the famous fusion reactions—deuterium reacting with deuterium—two of the most important ever observed. They will probably provide most of the world's energy after coal, oil, and uranium have been depleted. The ocean contains virtually unlimited supplies of deuterium, which can be separated from seawater with a very small fraction of the energy that is liberated when two deuterons interact. The only thing delaying the arrival of this utopia is what physicists like to call "a few engineering details."

The British group found that following such interactions the final products contained either a helium nucleus of mass 3 or a hydrogen nucleus of mass 3. These newly discovered nuclei could be observed only when they were moving at high speed, so no one knew what happened after they slowed down and picked up the one or two electrons they needed to convert themselves from ions to atoms. Two independent arguments convinced everyone that helium 3 was radioactive and that hydrogen 3—tritium, as it's now known—was stable. Stable tritium should therefore occur naturally in water along with ordinary hydrogen and deuterium. So the theory went.

Several mass spectroscopists searched for and reported finding stable tritium in ordinary hydrogen. By the time of my experiment, it was generally accepted that tritium was a stable and nonradioactive isotope of hydrogen but that helium 3 was unstable and radioactive. Bethe in his "Bible" illustrated the process of K-electron capture, yet to be discovered, by the expected decay of the heavier helium 3 into stable tritium. He calculated the lifetime of helium 3 against electron capture to be about five thousand years and concluded that it would no longer be found in nature, because it would long since have decayed.

Ernest Rutherford's last scientific paper concerned the mass-3 problem (he died suddenly in 1937 of peritonitis following a fall from a tree he was trimming in his garden on the Cambridge Backs, and the paper appeared in an issue of *Nature* just preceding his obituary). He used a sample of heavy water from the Norwegian electrolysis plant at Vemork that by 1937 had become the world's major source of supply. Thirteen thousand tons of ordinary water were electrolyzed down to about twelve grams of nearly pure heavy water, a volume reduction of one billion made possible by the plentiful hydroelectricity generated at the site. Rutherford gave this sample to Francis Aston, the inventor of the mass spectrograph, to test for traces of tritium. According to Rutherford, Aston found none at all.

Rutherford then tried an approach with which he was more familiar, bombarding the heavy water with deuterons to look for the long-range alpha particles that should come off when the deuterons interacted with any tritium nuclei in the target (we would now describe this as the D-T reaction; it powers the hydrogen bomb). He saw none, he reported; in the last three sentences he wrote in his fabulously successful life as an experimental physicist, he proposed that the two mass-3 isotopes could not be found, because they were used up too rapidly in nuclear reactions.

At home one evening, following my father's advice, I thoroughly reviewed the mass-3 problem. When I calculated how much of each isotope I could make in the 37-inch cyclotron by bombarding deuterium with deuterons for only an hour, I was impressed with how large the numbers were. If, after this short bombardment, I introduced the bombarded deuterium into the ion source of the 60-inch cyclotron and tuned the magnetic field to accelerate helium-3 ions, I could easily detect their presence with my thin ionization chamber. I was particularly excited at the prospect of detecting the presumably unstable isotope's radioactive half-life, if Bethe's estimate of its value turned out to be on the high side.

I asked Bob Cornog, then a graduate student, if he would like to work with me in exploring the fate of the mass-3 ions produced in D-D reactions. Bob hadn't yet done an experiment, and he was delighted. He agreed that we needed first to check the ion background in the 60-inch cyclotron. As far as I knew, no one had ever put an ionization chamber, which could detect individual ions, in a cyclotron beam while tuning the machine through a wide range of magnetic fields. A large background of junk ions in the helium-3 region would have made my proposed experiment impossible. We decided to take a look.

Bob helped me wheel my amplifier cabinet across the alleyway from the old, wooden, 37-inch cyclotron laboratory to the new, 60-inch cyclotron laboratory and then went off to throw the hammer in a track meet. I set the ionization chamber in front of the cyclotron's thin window and got it running. By the middle of the afternoon I was ready to ask the crew to operate the machine with its normal magnetic field so that I could observe both deuterons and ordinary helium ions. (The cyclotron was "filled" with helium.) The huge bursts of these ions paralyzed my amplifier, but it recovered quickly when the cyclotron's oscillator was turned off.

I next asked the crew to lower the magnetic field to about three-fourths of its normal value, the range where helium-3 ions should be accelerated. They spent half an hour moving the magnetic field around in this range. I glued my eyes to the oscilloscope screen, looking for pulses, the crew and I shouting back and forth. We couldn't find any junk ions being accelerated. This good news meant that the real experiment Bob and I planned to perform should be a success; there was nothing to keep us from detecting the helium-3 atoms we could certainly make across the alleyway.

What followed that afternoon was one of the finest moments of my scientific life. In the next few minutes I made an observation that not many physicists would have made or, having made, would have understood. As with every discovery, there was a substantial element of luck.

I asked the crew to take the magnetic field up to normal value again to make sure that my ionization chamber and amplifier were still working and then turn it off. The cyclotron operators proceeded to make a mistake. In good cyclotron practice the magnetic field is turned off only after power to the oscillator is cut. The operators had been running the magnetic field up and down all afternoon with the oscillator on, so now they cut the power to the magnets with the oscillator on. Knowing that the experiment was over when I shouted "Cut," I would normally have walked away from my apparatus. But for some reason I kept watching my oscilloscope that day.

Suddenly a burst of pulses registered and just as quickly disappeared as the magnetic field dropped through the helium-3 region. The magnet dropped slowly toward zero field because of eddy currents in its solid iron core. Soon I saw another huge burst of pulses as the field dropped through one-half the deuteron value, at which protons should be accelerated. When I saw the protons, I rushed to the control room and asked the operators to repeat running down the field. They did, and I always found helium 3 midway between ordinary helium ions and protons, but only when the magnetic field *was changing.* (And, as a magnet designer, I understood this strange behavior almost immediately.) With a set of calibrated aluminum foils,

I quickly measured the helium-3 range in air, confirming what I was seeing. The helium in the cyclotron came from deep wells in Oklahoma, where it had rested undisturbed for millions of years. Therefore helium 3 couldn't have the relatively short 5,000-year half-life Bethe had attributed to it. It was probably stable, as we now know it to be. Bob Cornog returned victorious in the hammer throw, and together we wrote a letter announcing our discovery to the *Physical Review.*

I'm embarrassed to say that it took me more than a week to realize that if helium 3 is stable then tritium must be radioactive. Fortunately no one else pointed out that obvious fact to me before I recognized it. When I told Bob about it, he took charge of the experiment and worked hard. We bombarded deuterium gas with deuterons in the 37-inch cyclotron and found it very radioactive. Most of that radioactivity came from impurities in the gas, but we passed the irradiated gas through activated charcoal to remove contaminants and diffused it through hot palladium to purify it. Hydrogen is the only gas known to diffuse through hot palladium; what came out the other side was pure hydrogen, and it was radioactive. Once again we reported our discovery to the *Physical Review.* Had Lord Rutherford lived another two years, the grand old man would, I'm confident, have sent along a note complimenting us on our work.

Willard Libby, who won the Nobel Prize in chemistry for the development of carbon-14 dating, told me a story once that demonstrates that Rutherford, like everyone else in the world before our discovery, believed that tritium was stable. Bill found himself in the Cavendish Laboratory shortly after the war and asked the curator of the Cavendish museum if he could locate Rutherford's sample of heavy water. When the curator did so, Bill held the sample near a Geiger counter, which went wild. That Rutherford never tried such a simple experiment proves to me that he was absolutely convinced of tritium's stability. (As Bill showed, the tritium in the sample had been made recently in the atmosphere by cosmic-ray bombardment.)

Hans Bethe's great compendium on nuclear physics was a handsome gift to his experimental colleagues. It deeply influenced my life. By combining Bethe's theoretical insights with the experimental imagination that has been my distinguishing quality as a physicist, I suddenly surged forward from the back of the pack of young researchers.

Our two short *Physical Review* letters had an enormous impact on science and even on world affairs. In the early days of the atomic-bomb program, the tritium work stimulated measurements of the likelihood of energetic fusion reactions between deuterium and tritium. They were found to occur at much lower ignition temperatures than anyone had guessed. The secret was well kept until the United States exploded the first hydrogen bomb, which contained deuterium and tritium. I helped open that Pandora's box. But if I hadn't made the discovery, someone else probably would have within a year; such is the nature of most science.

Another experiment during this period showed that I hadn't forgotten my first scientific love, optical spectroscopy, a field I had studied initially in Albert Michelson's Chicago laboratory. Michelson had spent years finding the narrowest spectral line, the red cadmium line, and demonstrated that it could be used as a universal standard of length. A heavy-element isotope would certainly give a narrower line if it could be separated from its fellow isotopes. But no one knew of a practical way to achieve such a separation.

In one of my meditation sessions I realized that mercury 198 could be made from gold 197 by neutron capture followed by beta decay. My student Jake Wiens accomplished that project, to the amusement of the popular press, which announced that two Berkeley scientists had reversed the alchemists' dream of transmuting base metals into gold. Similar amusement today would presumably greet a report of the first appendix transplant.

The mercury-198 green line was the universally accepted standard of length for many years after the war, but it was later supplanted by a line from isotopically separated krypton. At present, accurate short distances are measured in light-seconds just as astronomical distances are measured in light-years. Michelson, my first scientific hero, would be surprised to learn that there is no longer any need for measuring the velocity of light, a project to which he devoted much of his life; that velocity is now *defined* as one light-second per second.

Soon after my discoveries of the real properties of the two isotopes of mass 3, a stunning discovery was announced out of Germany. Then war came to Europe and changed all our lives.

FOUR

Fission

I LEARNED about the discovery of nuclear fission in the Berkeley campus barbershop one morning in late January 1939, while my hair was being cut. Buried on an inside page of the *San Francisco Chronicle* was a story from Washington reporting Niels Bohr's announcement that German chemists had split the uranium atom by bombarding it with neutrons. I stopped the barber in mid-snip and ran all the way to the Radiation Laboratory to spread the word. The first person I saw was my graduate student Phil Abelson. I knew the news would shock him. "I have something terribly important to tell you," I said. "I think you should lie down on the table." Phil sensed my seriousness and complied. I told him what I had read. He was stunned; he realized immediately, as I had before, that he was within days of making the same discovery himself.

In their classic series of papers reporting artificial radioactivity induced in a large number of chemical elements by neutron bombardment, Enrico Fermi and his colleagues at the Physics Institute in Rome in 1934 had noted that bombarding uranium with neutrons gave rise to a variety of radioactivities of different half-lives. They suspected, and tried to prove, that among those artificially created radioactivities were new elements beyond uranium (element 92 in the periodic table), transuranics never before seen on earth. "These two [chemical] reactions," they noted near the end of a paper mailed in July 1934, "appear to confirm the hypothesis that we have elements of atomic number higher than 92."

The Fermi group's proof involved chemical studies designed to show that the newly created radioactivities in uranium could not be attributed to any "nearby" chemical element. They compared the radioisotopes with elements of lesser atomic number, all the way down the periodic table to lead, atomic number 82, to which uranium eventually decays. Soon afterward a German chemist, Ida Noddack, published a critical demurrer: Fermi could not claim the discovery of new transuranium elements, she argued, until his unidentified radioisotopes had been compared with every element in the periodic table.

No one took Noddack seriously. The notion that uranium could turn into a lighter element in the middle of the periodic table under bombardment by nothing more energetic than thermal neutrons was self-evidently ridiculous; to do so, it would have to split, and the nucleus, we thought then, before Bohr elaborated the liquid-drop theory, was harder than the hardest rock, bound together by powerful forces—powerful enough to resist the electrical repulsions of all the protons. Everyone knew that the alpha particle—a helium nucleus, atomic number 4—was the largest chunk of nuclear material that could be chipped out of an atom. Nor was Noddack an entirely credible critic. She had shown her ability as a chemist by codiscovering, with her husband, the element rhenium, but the Noddacks had later announced the discovery of another element that proved to be mistaken, and they had continued rather shabbily to insist on the correctness of their work when the evidence demonstrated otherwise.

I was bothered at the time that the Fermi transuranics didn't fit the pattern of other radioactive elements. Instead of decaying, as we said, "downhill to the floor of the valley of stability," they decayed uphill, into a region that ought to be progressively more unstable. I had long been responsible for maintaining the big isotope chart that hung on the wall of the cyclotron control room, and every time I looked at it I was affronted to see the so-called transuranium elements decaying in the wrong direction. I knew something was off-key, as did everyone familiar with nuclear theory, but the correct explanation entailed such a radical departure from contemporary understanding that no one pursued the matter. It's a shame that Frau Noddack didn't follow up her own suggestion. She might have made, three years earlier, the epochal discovery Otto Hahn and Fritz Strassmann made at the Kaiser Wilhelm Institute for Chemistry in the Berlin suburb of Dahlem in 1938.

I had spent several evenings before then following my father's advice, trying to think of a way out of the madness I observed on the isotope chart. I would still feel like an idiot if Enrico Fermi, who was infinitely smarter than I was in such matters, had not also resisted Noddack's lead. Fermi received the Nobel Prize in 1938, only weeks before the Christmastime discovery of nuclear fission, "for his demonstrations of the existence of new radioactive elements produced by neutron irradiation and for his related discovery of nuclear reactions brought about by slow neutrons." He could have been awarded the prize for any of a dozen different theoretical or experimental discoveries in the course of his richly creative scientific life; it's ironic that the first citation on his Nobel diploma happens to record the only scientific mistake he made that I know of.

Phil Abelson was among my best graduate students. He had been trained in chemistry as an undergraduate, and he wondered about the chemistry of Fermi's transuranics. Since I had recently demonstrated that many radioisotopes emitted characteristic X-rays, Phil thought he might be able to determine the atomic number of the transuranium radioisotopes by measuring their X-ray spectra.

From the multitude of radioactivities produced by the bombardment of uranium with slow neutrons, he chose a radioisotope with a three-day half-life and separated it chemically from its cohorts. He showed that it gave off X-rays that were absorbed in aluminum the way the L X-rays of a transuranium element should be. He prepared to record its X-ray spectrum on a photographic plate. By measuring the position of the diffracted X-ray spectral lines, he could assign an atomic number to the isotope and determine if it was in fact what Fermi's group had proposed.

He was unlucky in the numerical values involved; on his first attempt the diffracted X-ray lines didn't hit his photographic plate. During the next week he would have changed his observation angle and obtained the telling pictures. They would have shown the simple K X-ray lines of a light element rather than the complex L X-ray lines of a heavier element, and the story of how Philip Abelson discovered fission would be history. But I bolted from the barber chair, and Phil wasn't given that extra week. Once he knew of fission, he quickly found that he was looking at iodine K X-rays; his isotope was tellurium, atomic number 52. With Phil hovering over me, I wrote out a telegram to the *Physical Review* for him, which he signed. The report "Cleavage of the Uranium Nucleus" appeared in the same issue with several other verifications of the fission discovery, one of them by Kenneth Green and me. Phil went on to a distinguished career, contributed vitally to the production of uranium 235 for the first atomic bomb, and served for many years as the much-respected editor of *Science.* His last work at Berkeley was as codiscoverer of neptunium, the first of the transuranium elements. His partner, Ed McMillan, won the Nobel Prize for that work, but, for reasons I never understood, Phil missed out on that high honor.

I also narrowly missed discovering fission. Like many of my colleagues around the world, I looked for long-range alpha particles coming from uranium bombarded by slow neutrons. Also like them —including Fermi's group—I covered the uranium with just enough aluminum foil to block the background of short-range alphas from uranium's natural radioactivity, thereby also blocking the fission fragments we would otherwise have seen. Years later I did discover the long-range alphas that are produced by fewer than 1 percent of fission events. I'm still surprised that I didn't find them at Berkeley in 1938. Had I done so, I would certainly also have quickly seen the large oscilloscope pulses due to fission fragments. I'm probably lucky to have missed the discovery of fission. I doubt if I had the maturity at twenty-seven to handle the burden of having made one of modern science's greatest discoveries. The implications of this finding disturbed Otto Hahn so profoundly, he reported later, that he seriously considered suicide.

As people arrived at the laboratory on that exciting late-January morning, we told them the news. Everyone found it hard to believe. I tracked down Robert Oppenheimer working with his entourage in his bullpen in LeConte Hall. He instantly pronounced the reaction impossible and proceeded to prove mathematically to everyone in the room that someone must have made a mistake. The next day Ken Green and I demonstrated the reaction. I invited Robert over to see the very small natural alpha-particle pulses on our oscilloscope and the tall, spiking fission pulses, twenty-five times larger. In less than fifteen minutes he not only agreed that the reaction was authentic but also speculated that in the process extra neutrons would boil off that could be used to split more uranium atoms and thereby generate power or make bombs. It was amazing to see how rapidly his mind worked, and he came to the right conclusions. His response demonstrated the scientific ethic at its best. When we proved that his previous position was untenable, he accepted the evidence with good grace, and without looking back he immediately turned to examining where the new knowledge might lead.

The extra neutrons—instantaneous "secondary" neutrons ejected in the fission process—soon became the object of a worldwide search. Frédéric Joliot, Lew Kowarski, and Hans von Halban in Paris and Fermi and Leo Szilard at Columbia University independently identified them in experiments of extraordinary difficulty. They had to find a few secondaries in the sea of neutrons that caused the uranium to fission, and they had no apparatus that could sort out fast neutrons from slow. A few years later, when Fermi had built the world's first nuclear reactor, at the University of Chicago, I amused myself by using its graphite thermal column to show the emission of secondary neutrons from uranium fission in an experiment that took half an hour from start to finish. I felt particularly stupid then because in 1939 I had a beautiful piece of apparatus—my neutron time-of-flight apparatus—that was equivalent in almost every way to a thermal column. If I had understood the importance of finding the secondaries, I could certainly have found them first.

I did decide for some reason one afternoon that I would look for them. I handled the work myself and didn't tell anyone what I was doing; but if I had continued for an hour or so I would have seen them. The boron trifluoride neutron detector in my ultracold neutron beam was set up just outside the cyclotron room on the staircase that John Lawrence's cancer patients used. I reasoned that I should look between cyclotron pulses when all fast neutrons had long ago passed by and only slow neutrons were present. If I then surrounded my counter with paraffin wax and surrounded the paraffin with a layer of cadmium, the counting rate at the detector would drop to zero. But if I put bottles of uranium compounds in the neutron beam close to the counter and the uranium gave off secondary neutrons, those particles would penetrate the cadmium, slow in the paraffin, and be detected: the counter would count them.

My reasoning was unimpeachable. I signed out several bottles of uranium oxide from the chemistry storeroom and made the necessary changes in my apparatus. Then I looked for the pulses from my counter—for about five minutes. When nothing turned up, I went back to what I had been doing a few hours before. (If only I'd known that Fermi and Joliot both had devoted all their time for the past several *months* to searching for these neutrons!) I could easily have increased the sensitivity of my experimental arrangement by a million times: by moving my counter nearer the cyclotron (x 20), by putting more uranium salt in the beam and more paraffin around the counter (x 500), by counting for an hour with and without uranium (x 100). I would have seen the secondary neutrons the same day. But I didn't; I was stupid. I didn't understand how important the experiment was, and I was too busy with several other experiments that I knew were important to see the importance for myself—I saw it only to the extent of trying a quickie experiment. Phil Abelson and I missed discovering fission, Emilio Segrè missed discovering the transuranium elements, and now I missed discovering the neutrons that accompany fission. The only consolation is that all three of these searches were pushing the state of the art. In compensation the Radiation Laboratory soon succeeded in discovering neptunium, plutonium, and the slow-neutron fission of plutonium.

Robert Oppenheimer's war work and postwar problems have made him a historic figure. Unfortunately, his relationship with Ernest Lawrence is often misunderstood, and his earlier work is slighted along the way.

Ernest and Robert were close personal friends. Ernest's second son was named for Robert Oppenheimer. The two men were distinctly different types, but they got along very well. Ernest came from a small town in South Dakota; Robert was brought up in a wealthy and cultured home in New York City. Robert knew foreign languages and enjoyed poetry, philosophy, and mysticism, none of which held any appeal for Ernest.

Before the war Ernest was already a man of the world. His contacts in politics and business were essential to his fund-raising. At that time Robert was unpractical and unworldly, a condition symbolized by his long, wild hair. But he proved to be a marvelous teacher, and the understanding of quantum mechanics that he imparted to his students is largely responsible for the way that important subject is taught even to beginning physics students today. He was the center of a loyal and extremely productive group of young theorists, as Ernest was of experimentalists.

Robert invested his spare time in those days in his students and in one other activity—left-wing politics. He was independently wealthy and often entertained students and colleagues at elegant San Francisco restaurants. Gerry and I were his guests once at the theater. He was a gracious host, but the production he chose for us, performed in a San Francisco union hall, was amateurishly devoted to the joys of the union movement. Such causes were important to Robert then. Ernest didn't share them. Once, when Robert wrote a notice on the laboratory blackboard announcing a cocktail party for relief of the Loyalists fighting the Fascists in the Spanish civil war, Ernest read it and simply erased it. Their friendship survived such differences. Ernest strongly supported Robert's appointment as director of the secret bomb-development laboratory established at Los Alamos in 1943, when some of Robert's closest friends were skeptical. "He couldn't run a hamburger stand," I heard one of them say. I was certainly surprised to learn of the appointment, but after working at Los Alamos later in the war I came to feel enormous admiration for the way Robert handled that terribly difficult job.

Robert Oppenheimer and his students were among the first to become interested in the gravitational collapse of stars. In that interest Robert was far ahead of his time. He didn't live to see two of his predictions sensationally verified. In 1939 he published two historic papers in the *Physical Review.* In the first he and George Volkoff predicted the existence of neutron stars—collapsed, superdense stars consisting almost entirely of neutrons—and developed the theory of these important stellar objects. In the second he and Hartland Snyder did the same for black holes. Physicists came to rate Robert as good but not first-rank and correctly predicted that he would never win a Nobel Prize. If he had lived until the late 1970s, when neutron stars —pulsars—were an established fact and the search for black holes was well under way, the Nobel committee would have recognized him, I believe, for his contributions to astrophysics.

For most Americans, World War II began the Sunday morning the Japanese attacked Pearl Harbor, December 7, 1941. For me it started on November 11, 1940, coincidentally the twenty-second anniversary of the armistice that ended World War I, when I left Berkeley for Cambridge, Massachusetts, to help with the development of radar. I worked on war-related projects for the next five years.

Alfred Loomis was a central figure in the events that led up to my departure; he became a close lifelong friend and a second father to me. His contributions to radar and to the project to build an atomic bomb have gone largely uncelebrated, an anonymity of which he would have approved but which hardly does him justice. He was the last of the great amateurs of science. I came to think of myself as, and he came to call me, one of his "other sons."

Alfred Loomis was born in New York City in 1887. His father, grandfather, and maternal uncle were all physicians; his uncle was also the father of Alfred's favorite older cousin, Henry Stimson, Herbert Hoover's secretary of state and Franklin Roosevelt's World War II secretary of war. Alfred attended Andover and Yale, where he excelled in mathematics, and graduated among the top ten in his class at Harvard Law School.

The Great War interrupted his early career as a corporate lawyer. He entered the Army officer corps with an encyclopedic understanding of modern artillery weapons, a hobby of his. He was distinguished by a wide-ranging mind and the ability to learn about a completely new field in a remarkably short time, and he applied that ability in this instance to making himself an expert on ordnance. As a result he was assigned to the Aberdeen Proving Ground, where he was soon put in charge of experimental research on exterior ballistics.

At Aberdeen he worked with some of the best physicists in the country. He became a physicist himself. After the war he made himself enormously wealthy through many kinds of business deals, especially in the financing of public utilities; some of his wealth he used to outfit a luxurious private physics laboratory in a huge stone mansion in Tuxedo Park, the wealthy suburban New York colony where he kept a home. Alfred and his business partner Landon Thorne, who was his brother-in-law, were prosperous enough to build an innovative J-class racing sloop for the America's Cup races at a time when even the Vanderbilts divided such expenses with syndicates of like-minded sportsmen. The partners owned, and kept as a personal and completely rustic riding and hunting preserve, Hilton Head Island.

Alfred's wealth was entirely facilitative. By the mid-1930s he had retired from Wall Street and was doing science full-time. He worked on ultrasonics and on accurate timekeeping, and he made significant contributions to the study of brain waves. He was the anonymous donor who paid page charges for impecunious Depression-era contributors to the *Physical Review.*

In 1939 his scientific interests changed drastically. He became deeply involved in Ernest Lawrence's projects. He also shifted the emphasis of his own laboratory from pure science to war-related technology by starting construction of a microwave radar system to detect airplanes. I was not surprised to meet Loomis in Berkeley on his first visit to the Radiation Laboratory in 1939; Francis Jenkins had spent a summer at Tuxedo Park as Loomis's guest and had talked in wide-eyed amazement about the fantastic laboratory there and the mysterious millionaire-physicist who owned it. Alfred came to Berkeley because he was excited by the work of Lawrence's laboratory. He had learned that Ernest was trying to raise money to build what became the 184-inch cyclotron and realized that with his contacts in the business and the philanthropic worlds he might be able to help. He spent several months in Berkeley.

The relationship that quickly developed between Alfred and Ernest was the equivalent of a perfect marriage. They were completely compatible, and their backgrounds and talents, though different, complemented each other well. Ernest had developed a new way of doing what came to be called big science. That development stemmed from his ebullient nature plus his scientific insight and his charisma; he was the most natural leader I've ever met. These characteristics attracted Alfred to him, and Alfred in turn introduced him to new and fascinating worlds.

I was impressed with the way Loomis sought out the younger members of the laboratory to learn about us and from us. I had never before discussed physics seriously with anyone as old as Alfred, and I was pleased that he enjoyed visiting with me after I had taught a freshman class and was sitting out my required office hour, waiting to talk with students who seldom came by. We found we were simpatico. He taught me the important lesson that a scientist can stay active as he grows older only by staying in touch with the youngest generation, the frontline soldiers.

I in turn introduced Alfred to more nuclear physics than he had known before, and we enjoyed discussing my research plans for the coming year. I remember one occasion when I mentioned in passing that because of the war in Europe the price of copper had risen to almost twice that of aluminum, volume for volume. Since aluminum has only 60 percent more specific resistivity than copper, I suggested that aluminum might now be the preferred metal for the magnet windings of the 184-inch cyclotron. It seemed obvious to me from elementary scaling laws that an aluminum coil would be larger but would cost less. I had completely forgotten the suggestion when, a few days later, Alfred showed me a long set of calculations based on several altered designs of the 184-inch cyclotron that proved my snap judgment wrong. I appreciated then for the first time the difference between the world of business, where a 20 percent decrease in cost is a major triumph, and the world of science, where nothing seems worth doing unless it promises an improvement by a factor of at least ten. I hadn't done the calculations, because they obviously didn't permit such large savings. Alfred, on the other hand, considered it worth a day or two of his time to see if he could cut the cost of the magnet windings by $50,000.

Loomis returned to Berkeley again early in 1940. By then he had helped Ernest raise $2.5 million from the Rockefeller Foundation for the 184-inch cyclotron. I asked him what he was going to do next. He said he was going to help Enrico Fermi find the necessary funding to build a nuclear chain reactor. But Loomis's immediate involvement in reactors was cut short in the summer of 1940 by the dramatic appearance in Washington of the British Tizard mission.

Henry Tizard, an Oxford chemist by training, had been the driving force behind British radar development as chairman of the famous Tizard committee. Tizard first became concerned with the defense of Britain against air attack in about 1934 and organized his committee of scientists to see what could be done. They concluded that whatever systems might be available—fighter aircraft, antiaircraft guns, or mere shelters—an absolute necessity was at least twenty minutes' warning before enemy bombers arrived. The committee then distributed its findings among British scientists, asking for help.

One who responded was a radiophysicist, Robert Watson-Watt. He reported that he had been measuring the height of the ionosphere for several years by sending short pulses of radio waves upward as high as one hundred miles, bouncing them off the ionosphere and measuring the time it took for them to return to the ground. The ionosphere was ubiquitous and therefore an easy target, but with more power and larger antennas, he believed, it should be possible to detect bombers by a similar process while they were still perhaps sixty or seventy miles away. That would give twenty minutes' warning.

Watson-Watt undoubtedly wrote out the basic radar equation in his reply to the Tizard committee; it's derived and analyzed in the first chapter of every book on radar I've ever read. It demonstrates that for an aircraft of a given size, the maximum range of detection varies directly with the linear size of the antenna but only as the fourth root of the *average* signal power—not the *peak* power as one might guess. The fourth-root dependence means that increasing the range by increasing the signal power is expensive; one needs sixteen times as much power simply to double the range. Range also increases slowly with decreasing wavelength, but Watson-Watt didn't have that option. There weren't any high-powered transmitters available then for wavelengths shorter than the ten meters he was using in his ionosphere research. So he proposed to use very large antennas, more than a hundred feet tall. By locating them on the cliffs overlooking the English Channel, he gained the extra height that gave these "Lloyd's Mirror" antenna systems such surprising vertical resolution that radar operators were able to determine within a few hundred feet how high an incoming bomber was flying.

The outcome of the Tizard–Watson-Watt collaboration was the emplacement before the war began of the Chain-Home radar system that was essential to the defeat of the German Luftwaffe in one of history's decisive battles, the Battle of Britain. Neither the fighter pilots nor the radar operators could have done the job by themselves; together they were unbeatable. To engage the bombers, the fighters had first to climb to altitude; that was how they used their warning time. Without that warning time they could have been effective only if they patrolled continuously on station at altitude, and the British did not have enough aircraft for such continuous patrol. Finding the fighters at altitude when their bombers arrived and not understanding the effectiveness of the Chain-Home system, the Germans in consequence seriously overestimated British fighter strength. I heard later from Royal Air Force officers that the number of serviceable Spitfires and Hurricanes left when the Luftwaffe broke off the battle was under twenty. I haven't checked the number. It was certainly small.

After such a triumph of foresight, Henry Tizard might have seemed the obvious choice to be the British government's principal science adviser. For a time he was, but when Winston Churchill took office as prime minister he brought along a favorite science adviser of his own, Frederick Lindemann, a wealthy physicist who directed the Clarendon Laboratory at Oxford. The new favorite soon shouldered Tizard aside. Tizard's committee had lost much of its influence by the summer of 1940. Tizard was still the man of choice for a vital mission to the United States. That mission would enlist the help of Britain's unofficial ally—the United States was still officially neutral at the time—to develop and build the new technology needed to meet the military requirements of a war that had become dependent on it to a degree that remained largely unappreciated by our military, industrial, and scientific establishment. (Radar, for example, had been invented independently in the United States by the Navy and the Army. The military shrouded the work in such excessive secrecy that no outsiders learned of it. Since the outsiders were the real professionals in radio engineering, they were the ones who could have developed American radar into the useful, integrated system that the insiders failed to achieve. Pearl Harbor demonstrated the dismal state of U.S. radar a year and a half after the Tizard mission showed us the importance of integrating radar information with military command and control. The technical side of radar worked at Pearl Harbor, but the officers responsible for interpreting its information hadn't been well trained and chose to ignore the clear warning signals.)

After they developed the Chain-Home radar system, Watson-Watt and his young colleagues turned their attention to airborne radar. A second generation of VHF radar in the 1.5-meter band could be fitted onto aircraft to turn them into night fighters and antisubmarine patrols. Everyone agreed that microwave radar in the 10-centimeter band would be vastly superior to the 1.5-meter equipment then available. But the chance that a powerful generator of such pulsed microwaves could be developed seemed small.

Then John Randall and Henry Boot, physicists working with Mark Oliphant in the laboratory he directed at the University of Birmingham, invented the cavity magnetron, the world's first high-powered source of microwaves. It was a remarkable invention. By comparison, my research adviser at the University of Chicago in the mid-1930s was generating microwaves with a simple magnetron tube that had a power output of less than one milliwatt, 1/1000 of a watt. RCA produced a radar set in 1934 with the unprecedented power of thirteen watts. Randall and Boot's first cavity magnetron yielded fifty kilowatts, and within about two years its direct descendants were generating pulses of more than a megawatt. A sudden improvement by a factor of three thousand may not surprise physicists, but it is almost unheard of in engineering. If automobiles had been similarly improved, modern cars would cost about a dollar and go a thousand miles on a gallon of gas. We were correspondingly awed by the cavity magnetron. Suddenly it was clear that microwave radar was there for the asking. But Britain had no scientists and engineers to spare.

Everyone was working at breakneck speed on the immediate challenges of a desperate war that could be lost any day.

In a great and successful gamble, Winston Churchill made the decision to share all his country's technical secrets with the United States. Those secrets included more than the cavity magnetron. They included new explosives; magnetic mines; first efforts to design a radio-based proximity fuse that would explode antiaircraft shells in lethal proximity to aircraft and thereby spare gunners the necessity of a direct hit; and, most significant in the long run, evidence of the real possibility of building an atomic bomb. Tizard brought this information over with a crew of talented scientists and with plans and models stored in a black-enameled metal steamer trunk, the original black box. The U.S. military was reluctant at first to share its secrets in kind, but the mission was finally a success.

John Cockcroft, Ernest Lawrence's old friend, came over to talk about radar (and the atomic bomb), as did Watson-Watt's young protégé Edward ("Taffy") Bowen. Alfred Loomis was included in the briefings not only because of his unique position in the scientific establishment but also because he had built one of the few microwave radar sets then in existence in the United States. It was what we would now call a police radar, which measures velocity but can measure distance only crudely. Alfred had been excited by the military potential of these continuous-wave radars. When he learned about the cavity magnetron, with its enormous peak power, he instantly recognized the superiority of pulsed radar over the continuous variety and pushed the newly formed National Defense Research Committee (NDRC) into full-time pulsed-radar development. The radar experts with the Tizard mission spent a week as Loomis's guests at Tuxedo Park, working with the newly established microwave committee of the NDRC, of which Alfred was chairman and Ernest Lawrence a member. The committee decided to establish a Radiation Laboratory at MIT. (The name was a cover intended to mislead the curious, who might imagine we were working on something so impractical as an atomic bomb.) Loomis would arrange to equip the laboratory with the necessary hardware. Lawrence agreed to find the laboratory staff. On the advice of the Tizard people, most of us would be young nuclear physicists; the British visitors said that, from their experience, we might prove more adaptable to a new set of ground rules than would radio engineers.

Ernest persuaded the University of Rochester's Lee DuBridge to direct the new MIT laboratory. That was a fortunate choice; Lee was a great director. Lawrence traveled all over the country recruiting his former students and their colleagues from the cyclotron laboratories they had modeled after Berkeley. He didn't spare his own. Ed McMillan and I were first on his list to be recruited after Lee DuBridge. And that's how I came to make my early entry into the war.

My son Walter was born on October 3. We named him for his grandfather. Gerry and I had just moved into a charming house with a spectacular panoramic view of San Francisco Bay. Before Ernest recruited me for MIT, we thought we'd be settling down for a long stay. How inaccurate that scenario was is suggested by the nine different homes Walt knew in the first six years of his life.

I might have felt disappointed at leaving nuclear physics, but radar work was even more exciting and exhausting. In a few weeks I felt sorry for my MIT colleagues who were stuck in nuclear physics when they could have been working on radar. Any kind of science can be compelling if the people one most admires are interested in one's results. I was in the midst of my most productive years as a working scientist. At the time I didn't think about the physics I might have done had I stayed at Berkeley. That loss can never be restored. The war years demanded other and necessary assignments: the development of useful radar systems and work at Chicago and Los Alamos on the atomic bomb.

FIVE

Radar

ED McMILLAN, four years older than I, received his Ph.D. from Caltech in 1932 and came to Berkeley as a National Research Fellow. His talent for experimental physics was obvious, but almost uniquely in our company he also knew theory. Most experimentalists then thought of themselves as high-class plumbers and asked their theoretical friends for help whenever they had a difficult quantum mechanics problem to solve. Ed knew chemistry as well (the field in which he eventually won his Nobel Prize) and made an outstanding contribution to electrical engineering by designing the innovative constant-current power supply for the 60-inch cyclotron.

He and I had worked together in bringing the beam out from the 37-inch cyclotron and assembling the 60-inch. We were the only Berkeley Radiation Laboratory physicists given faculty appointments before the war. It wasn't surprising that Ernest would send us off to MIT together.

The MIT Radiation Laboratory was to open on November 15. We had only a few weeks to organize research programs to be carried out in our absence (we anticipated returning at Christmas but realized independently long before then that development was progressing too rapidly for us to leave MIT; we'd have returned hopelessly out of date). Geraldine's mother came to baby-sit just before her daughter was to have our baby and several weeks before I went to MIT. In the meantime my graduate students baby-sat my apparatus. I spent my last days in California writing two coauthored letters to the *Physical Review.* Since the faculty substitutes we found thought it would be helpful to have taught a course in the Berkeley physics department, my classroom duties were well covered.

Ed and I departed Berkeley on the November 11 afternoon train. We had often seen Ernest off, but the bon voyage party that afternoon was a first for us. Besides some of our students, it included Don Cooksey and Ernest himself. Robert Oppenheimer gave each of us a bottle of whiskey. Robert normally gave friends books to read on transcontinental trains but had recently been discouraged from making such gestures of improvement. He had seen the deeply reserved theoretical physicist Paul Dirac off on a visit to Japan. Dirac politely refused Robert's two proffered books; reading books, the Cambridge theoretician announced gravely, "interfered with thought."

In Boston, Ed and I checked into the Commander Hotel, our home for the next several months. I was visiting MIT for the first time and found it fascinating. The school's buildings were not named but numbered; the Radiation Laboratory was establishing itself in half a dozen moderate-sized rooms behind a door labeled 4–133. The space was split between two floors connected by an internal spiral staircase. Lee DuBridge, a talented physicist who had been actively experimenting at Rochester, moved into the small director's office and abandoned his former way of life for administration. In my three years at the laboratory, I never heard Lee make a single technical suggestion. I greatly admired his obvious restraint and the full devotion of his talent and energies to his important administrative job.

The basis of radar is a transmitted electromagnetic pulse reflected by an object of interest to produce a pulse at a receiver at some later time, the elapsed time serving to measure the distance. The British had found that nuclear physicists had greater aptitude for pulsed radar development because we made our living with electronic pulses from ionization chambers and Geiger counters. Radio and electronic engineers, who were more familiar with the quasi-sinusoidal signals used in audio or radio-frequency engineering, abhorred pulses; such signals usually indicated trouble—lightning flashes or sparking.

The first weeks at the laboratory were like a family reunion. Ernest had recruited, among others, I. I. Rabi, Ken Bainbridge, Norman Ramsey, S. N. Van Voorhis, Louis Turner, Milt White, Alex Allen, Vic Neher, Ivan Getting, Ernie Pollard, Jim Lawson, Curry Street, and Ed Purcell. All of us began recruiting former colleagues and the laboratory grew exponentially. I tracked down George Comstock, Howard Doolittle, and Andy Longacre, fellow graduate students in my Chicago days. They had all accepted dead-end academic jobs in small colleges during the Depression; if the war hadn't come along, they would have retired eventually from those duties. Instead, George and Howard found responsible and high-paying positions in industry at the end of the war, and Andy went to Illinois as a professor of physics.

But the most important person to join the laboratory was Taffy Bowen, who, we later realized, had designed the airborne microwave radar set we were to assemble and test. From Taffy we learned that the invention and development of radar equipment was only a small part of the Watson-Watt team's triumph. Because radar was so extraordinarily valuable, the RAF built its organizational structure around it, chains of radar sites tied to Fighter Command headquarters and to individual RAF aerodromes.

Taffy also filled us in on the development of night-fighter radar, a 200 megahertz system—now TV channels 7 to 13—that the new microwave radar we were to assemble and test was meant to replace. The British night-fighter radars had only a short range because their antennas were insufficiently directional. Radio waves reflected from the ground blinded the signal receiver at distances greater than its altitude. Such limitations dictated the development of a new kind of ground-based radar—ground control of intercept (GCI). This system operated by using the low-frequency chain stations first to pick up incoming bombers. The high-frequency GCI radars would phase into operation next with their more-accurate but shorter-range beams, detecting both the bomber and the defensive night fighter ascending to meet it. The ground-based radar operator then used the return signals from both planes to instruct the fighter pilot which course to fly to close on the bomber. The airborne radar in turn led the fighter to within visual range. Once visual contact was established, the pilot used his guns to bring the enemy down. The complexity of these sequenced operations made it clear to everyone that night fighters needed more sharply defined radar beams, and that requirement dictated the use of microwaves. So the invention of the cavity magnetron by Randall and Boot signaled the coming of age of airborne radar.

I vividly remember Air Chief Marshall Hugh Dowding, who led the RAF Fighter Command during the Battle of Britain, talking to our little band of nuclear physicists turned fledgling radar experts. Dowding had none of the charisma that Americans expect in their generals; he was all business, and as he spoke I could understand how he had earned the nickname Stuffy. American fighter planes in 1940 were still armed with a single machine gun firing through the propeller blades, World War I style. Dowding had armed his fighters with six or eight machine guns mounted in the leading edges of the wings outboard of the propeller arc, with their paths of fire converging a few hundred feet ahead. A sixteen-millimeter movie camera automatically began filming whenever the fighter pilot pressed his trigger. Dowding concluded his fascinating talk by showing us his secret gun camera films of British aircraft destroying German fighters and bombers. He fired our enthusiasm; we were eager to get on with the job of building a really modern night-fighter radar system.

We divided ourselves into groups to match the components that would soon arrive from industry. Ed and I joined I. I. Rabi's magnetron group but maintained our scientific independence by frequently crossing from one group to another. We soon made ourselves expert in most aspects of the early work. Later on, of course, each of us picked out a particular area and specialized. I was excited when several cavity magnetrons arrived. Bell Labs reproduced them within two months of the Tizard mission's visit. They marked our first experience with secret hardware, and we fretted about security; we kept them in a small safe in Lee DuBridge's office.

Not long after they arrived, Ed and I took one of the magnetrons to Raytheon to see if we could interest Percy Spencer in manufacturing them; he was famous in the vacuum-tube industry for being able to make anything. Our visit proved to be the turning point in Raytheon's fortunes. The company was then one of the smaller and less profitable in the radio business. Percy invented some clever magnetron production methods. They made a foundation on which the company subsequently built its fortunes. Not long after this visit the president of Raytheon, Larry Marshall, invited Ed and me to Thanksgiving dinner. It never occurred to us that we had done the company an enormous favor.

When the microwave committee met at MIT in early December,

Ernest and Alfred decided that I should be appointed coordinator responsible for producing the first pulsed microwave radar set. I don't know why they picked me. I had never coordinated production before or even demonstrated any unusual talent for keeping a program on schedule. Many of my projects have finished on schedule, but I think that result reflects a talent for attracting people to my teams who are good at scheduling. Ernest and Alfred also decided that I would be responsible for producing a limited number of the early-model radar sets when the laboratory began manufacturing them. That decision was certainly one of the worst either of them ever made. If other duties hadn't shunted me away, I would certainly have fallen flat on my face. On three other occasions in the years we worked together, Ernest seriously proposed me for administrative jobs for which I had little, if any, talent. Fortunately I was able to avoid them all.

As coordinator I roamed the laboratory learning the details of the components that made up the radar system. I quickly rediscovered and put to good use the enthusiasm I had noticed earlier among physicists for explaining their work to the nearest interested listener.

One essential subassembly, a parabolic dish three feet in diameter with a two-inch dipole antenna at its focus, needed a component no one had yet invented. Since space is limited in the nose of an aircraft, the antenna had to serve for both transmitting and receiving. The transmitter emitted a powerful, fifty-kilowatt microsecond pulse. Despite the paralysis such a blast induced, the receiver had to detect reflected signals a few microseconds later with a trillionth that strength. Separate antennas for transmitting and receiving would have solved our problem; such systems are called bistatic. But to make our monostatic system work we needed a transmitting-receiving switch, which no one knew how to build. Fortunately Jim Lawson found a temporary solution using a klystron amplifier in the line from the antenna to the receiver input. The klystron recovered quickly from the fifty-kilowatt pulse and acted to prevent it from reaching the sensitive germanium crystal in the receiver, which it otherwise would have burned out.

Jim had a strong amateur radio background. The rest of us, by contrast, had learned the few RF techniques we knew from building and operating cyclotrons. If we had been paid in proportion to our contributions to the success of the first microwave radar program, Jim Lawson would have earned more than half the monthly payroll.

The other original lab member with extensive radio engineering experience was our Berkeley colleague Winfield Salisbury. Working on the first system one day, we saw large intermittent noise in the display of the receiver output oscilloscope. We nuclear physicists had absolutely no idea what to do. Win calmly asked if anyone had a pair of earphones. An MIT engineer found a set in a student laboratory. When Win hooked the earphones to the radio receiver, he heard a voice—"Hello CQ, CQ, hello CQ"—an amateur radio operator announcing himself on the air. With proper shielding the formerly mysterious noise disappeared, and we were back in the radar business again.

The first radar signals we detected we bounced off the dome of the Christian Science mother church, in nearby Boston. With that system set up and working on the MIT roof, I next designed and built with my own hands a wooden mock-up of the bombardier's compartment of the B–18 bomber. I installed a second radar system in the mock-up to serve as a test arrangement until the B–18 arrived. Our B–18 was the first airplane ever fitted with a radome. We made it of Plexiglas for transparency to microwaves and for aerodynamic strength. Such devices on commercial aircraft are today used to detect the intense concentrations of rain associated with dangerous thunderstorms.

Ed McMillan was in charge of the B–18 radar set. It was a great day for the Radiation Laboratory when he pronounced the unit ready to fly after extensive ground—roof, really—tests. Taffy Bowen and I flew with Ed as observers. Proceeding eastward from Boston, we were excited to see Cape Cod on the radar screen and surprised at how clearly large ships showed up. Taffy said that he would like to try to find a submarine. We asked the pilot to head for New London. As we flew low over Long Island Sound, several surfaced submarines reflected large signals. That was our most valuable observation, since airborne microwave radar turned out to be much more important for submarine detection than for night fighting. After the Luftwaffe retired from the Battle of Britain, German bombers had only a nuisance value. The German submarine campaign against Allied shipping, on the other hand, could have starved the British to the point of surrender. The threat was contained in the nick of time by an intensive campaign of antisubmarine warfare. Airborne radar was one of the keys to its success. (The other essential key, which none of us learned about until years after the war was over, was the British ability to decode the radio messages to and from German submarines.)

Nuclear submarines are true submarines capable of operating submerged for months at a time. By contrast, the submarines of both world wars were surface vessels that could submerge occasionally for short distances. They operated underwater on storage batteries that were quickly drained and had to be recharged by diesel engines in long stays on the surface.

Before airborne radar, a surfaced German submarine was visible from only a few miles away. RAF patrol planes rarely saw one. When the patrols got the 200-megahertz radar sets, submarine sightings and sinkings rose abruptly and stayed at a high level for about a year. Then the Germans caught on. They equipped their submarines with simple receivers that picked up the radar pulses and allowed submarine captains to submerge their boats until a patrol passed. The introduction of ten-centimeter radar led to an even higher level of submarine destruction. The Germans learned to detect those microwaves, and sightings returned to the very low visual level. Three-centimeter radar restored the Allied advantage. Eventually the Germans countered with three-centimeter listening devices.

I invented a system in 1941 that reactivated the three-centimeter antisubmarine campaign and led to the destruction of a large number of German U-boats. It was known as VIXEN because it foxed the submarine listening sets. It was based on two observations.

First, even a low-quality submarine receiver could detect a patrol plane long before the patrol could detect the submarine—the submarine receiver picked up the powerful direct beam, but the plane had to monitor the small, scattered U-boat reflection. In fact, the submarine waited on the surface and watched the way the signal from the aircraft increased in strength. If the signal increased slowly, then the patrol track was off to one side; that meant the submarine wouldn't be observed. If, on the other hand, the signal increased rapidly, the U-boat would dive before the patrol came close enough to attack.

Second, the different ways the signals varied with distance as seen by the submarine and the patrol plane appeared to me to be useful. The radar signal at the submarine depended on the inverse square of the range, while the returned signal at the airplane depended on the inverse fourth power of the range. So if a patrol on making contact with a submarine decreased its radar transmitter pulse power proportionally to the cube of the range, the submarine would misread the signal.

A patrol equipped with an ordinary radar set could see a surfaced submarine at twenty miles with minimum reflected-signal strength. The signal strength at a range of ten miles was sixteen times larger. A submarine observed a large signal at the first distance that increased fourfold at the lesser distance. That information alerted the submarine commander to dive.

But for a VIXEN-equipped patrol plane the signals at twenty and ten miles would change quite differently: the detected signal at the airplane would double, while the signal at the submarine would be cut in half. Approaching the submarine, the patrol would read an ever-increasing signal strength and be guided by it. The U-boat at the same time would hear an ever-decreasing signal, convincing it that the plane was going away. It would therefore stay on the surface, where the intense noise of its diesel engines would mask the sound of the approaching patrol. The patrol, easily tracking the submarine by now, would drop low over the ocean, its battery of forward-firing rockets at the ready. At quarter-mile range it would switch on its high-intensity Leigh lights, and the pilot would take aim. The system worked and saved millions of tons of shipping.

I was scheduled to visit the British radar laboratories in the spring of 1941 to bring them up to date on our microwave radar effort. Ken Bainbridge had been the first Radiation Laboratory scientist to visit England; I was to be the second. He flew the Pan American Clipper across the South Atlantic from Brazil to Dakar and up to Lisbon, which was neutral but swarming with spies. The flight on from Lisbon to Southampton was dangerous, the risk somewhat reduced by flying at night. Ken briefed me on the people I would meet and the subjects they would want to discuss. He also brought back news of the British atomic-bomb project, which was considerably ahead of ours though still at laboratory scale. Neither of us was cleared for such information, but our connections kept us informed. The British wanted American participation and talked freely to Ken to pass along the message.

I arranged to be immunized, collected my passport, scheduled my Clipper reservation. The National Defense Research Committee insured my life for a surprisingly large sum. I'd been having chest pains, however, and went to an internist for a checkup. My electrocardiogram was normal. My gallbladder history aroused suspicion, and an X-ray revealed stones. The internist could have given me morphine to tide me over an attack en route, but he was wary of the surgical risk in Britain's crowded wartime hospitals and recommended I have surgery before I go. I was terribly disappointed to be denied my British adventure. He assured me I could fly the North Atlantic later in the summer if I had the operation.

Geraldine and I packed our few possessions and drove to Rochester, Minnesota. It was a pleasant drive across New England with overnight ferryboats on Lake Erie and Lake Michigan. Our son Walt, six months old, slept most of the way in his basket. In the Midwest we discovered that almost everyone believed that the war in Europe was none of our business.

The operation, by Dr. Will Mayo's son-in-law Waltman Walters, was a success, but I woke severely nauseated from the ether. I continued in distress for the next two weeks, bedfast under sedation. That treatment reduced blood flow, and I developed phlebitis, a dangerous tendency to form blood clots. Thereafter I wore heavy elastic stockings and took daily doses of an anticlotting drug.

I spent most of the summer recuperating from the phlebitis in Minnesota. With a physician who was a family friend, I devised a more sensitive method for detecting gallstones; it involved X-ray films and the "critical absorption method" I had used to discover K-electron capture. We experimented with a phantom, a jar of water to which we added bones and a rubber balloon filled with gallstones and an iodine dye. Our new method was more sensitive, we concluded, but the gain in sensitivity was not worth the extra trouble.

Gerry and I rented a new two-story house in Belmont when we returned to MIT and had our furniture shipped from Berkeley. Wheeler Loomis—he was not related to Alfred—had joined the Radiation Lab as director of personnel and had made the happy decision to pay his formerly poverty-stricken young physicists salaries competitive with those of industrial scientists and engineers. We felt incomparably more prosperous than we had ever dreamed of being, and our improved morale improved our output. Wheeler's decision was proper. We really were working as industrial scientists now. The laboratory to which I returned was different from the one I had left. I had been away for nearly half its productive life. In the meantime activities and staff had grown exponentially. It took me a while to catch up.

At the outset the microwave committee had given the Radiation Laboratory two other assignments besides airborne radar: a pulsed navigation system that Alfred Loomis had invented called Loran and a ground-based antiaircraft fire-control radar that materialized as the SCR–584. In my absence Louis Ridenour had mounted an operating 584 prototype on the MIT roof. The British version of this system used two operators turning hand cranks to equalize the signal returned through two pairs of off-center antennas, the radar equivalent of traditional optical sighting. Louie made the wise, though not unopposed, decision to go fully automatic. He had studied industrial servomechanisms and found an appropriate system. Ivan Getting designed the SCR–584 so that its six-foot antenna would continuously track an airplane, feeding elevation, azimuth angle (that is, compass direction), and range to an analog computer that would predict the airplane's location by the time an antiaircraft shell arrived. The concept was wholly new and fabulously successful. Several hundred 584s saw service during the war; the device was one of the three scientific components that defeated the German V–1 buzz bombs (the other two were the proximity fuse developed by Ernest Lawrence's boyhood friend Merle Tuve at the Department of Terrestrial Magnetism of the Carnegie Institution of Washington and an analog computer from Bell Labs that pointed the guns).

One of my most valuable ideas came to me while I was watching the 584 prototype track an airplane. Since the Army and the Navy had not yet become sufficiently interested in the project to supply military aircraft for tracking, Louie had arranged for Dave Griggs, a Harvard geophysicist and amateur pilot, to fly as target. Dave wasn't a member of the laboratory staff but was later cleared. He found our work so much more interesting than flying around in circles that he became one of our most successful radar designers.

What occurred to me was that if a radar could continuously and automatically track an enemy aircraft accurately enough to shoot it down, the same information should be adequate to guide a friendly pilot to a safe landing in bad weather. This observation led to the development of ground-controlled approach (GCA). I thought it would be easy to convert a 584 into a blind-landing radar. It wasn't. I had to design a completely new system, including the first microwave phased-array antenna, to make GCA work.

The SCR–584 prototype on the roof was the first of its kind. The second, designated XT–1 and truck mounted, was finished in November 1941 and sent to Fort Monroe, Virginia, for evaluation. Evaluation would be finished in the spring, after which we could convert the unit for GCA. In a test one night at Logan Airport, Dave Griggs demonstrated that the XT–1 could track an aircraft on a standard glide path three degrees above the horizon beginning several miles from touchdown. Dave attached a bore-sighted movie camera and telescope to the XT–1 antenna mount and showed that he could see the aircraft throughout its approach. Further testing would have revealed serious deficiencies when the system was pointed at targets close to the horizon, and we probably would have abandoned the idea entirely.

The Griggs report and the early availability of the XT–1 for GCA trials set my group to working hard. We had to prove first of all that a pilot could land solely on verbal instructions from the ground. In the Logan National Guard hangar our blindfolded "pilot" walked a line on the floor while a "controller" observed his deviation and instructed him how to correct his path. The pilot could walk quickly along the line with only occasional instruction, which was encouraging.

Next we designed and built two simple optical sights—theodolites—through which we could observe an airplane landing and read its angles of elevation and azimuth. The system was wasteful of manpower, using one observer to measure azimuth, another to measure elevation, another to measure range with a primitive radar set, and a fourth, a controller, to talk to the pilot. We built a device, called the director, that accepted the two measured angles and the range and fed out three readings to the controller: the distance to touchdown in miles and the horizontal and vertical departures in feet from the plane's ideal glide path.

By now the laboratory had acquired a Navy test pilot, Aviation Chief Machinist Mate Bruce Griffin, and we invited him to the laboratory to see what all his mysterious flying had been about. He was happy to find real people at the other end of the radio communications link and fascinated by the radar and our dreams for GCA. He joined us as we were designing the director, and he contributed many useful ideas.

Bruce suggested one day that he give me lessons in the Link instrument trainer so that I would understand better the problems pilots faced in flying on instruments and eventually learn to land by GCA myself. I had thoroughly enjoyed my brief flying experience while a graduate student and enthusiastically accepted his offer.

I drove to Squantum Naval Air Station several times a week to prepare for the Navy instrument pilot's exam. Bruce was a born teacher. Accustomed to teaching fighter pilots, who flew by the seat of their pants and had little interest in instruments, he was surprised by my rapid progress. I assured him it would be common among nuclear physicists who had logged many hours flying a cyclotron from its control table. The Link trainer had a light-tight hatch cover to exclude exterior sensory cues. The most difficult part of instrument flying under such conditions is the response lag. Pull back on the stick to make the airplane go up, and the altimeter needle holds steady for a time and then suddenly indicates a rapidly accelerating climb. Pushing the stick forward to level off doesn't immediately stop the altimeter needle from its upward winding, so the plane overshoots. The cyclotron's magnetic field had a similar lag because of the magnet's inductance. Some months later Bruce and I invited the lab group leaders to Squantum for an accelerated Link trainer course. Just as I had predicted, the physicists who had operated cyclotrons could almost immediately compensate for instrument lag and overshoot. (Decades afterward I had the pleasure of attempting three moon landings in the Lunar Module Trainer in Houston under the watchful eye of my astronaut friend, Joe Allen, who gave up some of his practice time to see what I could do. I crashed on my first two attempts but greased it in on my third try.)

In March 1942 our small team drove to Quonset Point Naval Air Station, on the eastern shore of Narragansett Bay, with our two theodolites, our director, and our small range-only radar. Bruce Griffin flew down from Squantum in the Duck, his odd-looking Grumman J2F amphibian. During several weeks of work we gained confidence in our unorthodox solution to the blind-landing problem. We improvised communications procedures as we went along, and after many landings under conditions of good visibility Bruce began to fly under the hood to lower and lower elevations before making a visual approach. His radioman acted as check pilot when he was under the hood. One day Bruce announced that he had flown under the hood all the way to touchdown. We cheered the first complete landing on GCA. The fact that it had been made in perfect weather with angular information derived from optical theodolites failed to caution us. We were confident that we would enjoy the same success with the XT–1.

Finally, in May 1942, the Navy invited us to try out our system at a small field at Oceania Naval Air Station, near Norfolk. Ivan Getting and his friends came along to operate the XT–1. The Navy made available an SNJ advanced trainer. Bruce was delighted. Most Navy pilots considered the SNJ the best plane ever built. It was ideal for aerobatics; when Bruce brushed up on the loops, rolls, split-S turns, and Immelmanns he couldn't attempt in his ungainly Duck, I went along in the rear seat for the excitement of the ride.

Our GCA tests were disastrous. The XT–1, which had worked so beautifully in its one previous test against a low-flying aircraft, now went spastic. When Bruce flew toward the truck, the antenna would alternately point at the plane and at its mirror image three degrees below the surface of the runway. We tried any number of tricks to keep the radar from locking in on the reflected image. It was soon obvious that the SCR–584 would never work in an operational GCA system. We still hoped to practice radar-controlled landings. We moved our equipment to a nearby Coast Guard air station where we thought reflection might not be such a problem, but the radar worked no better. We returned to MIT thoroughly discouraged. Our most sensible option was to give up and move on to one of the many other problems radar could help solve. We'd fallen in love with the GCA talk-down technique, however, and no other blind-landing system would be operational for years.

The crucial breakthrough came at dinner with Alfred Loomis and his wife in their suite at the Ritz Carlton in Boston. Alfred and I analyzed the Quonset Point and Oceania events in detail. He did an amazing job of restoring my morale, which had hit a new low. "We both know that GCA is the only way planes will be blind-landed in this war," he told me, "so we have to find some way to make it work. I don't want you to go home tonight until we're both satisfied that you've come up with a design that will do the job." I'm convinced that without Loomis's ultimatum that night World War II would have seen no effective blind-landing system. I would have immersed myself in other interesting projects to forget my disappointment and embarrassment. Many lives would have been lost unnecessarily.

We both contributed ideas. The antenna configurations we devised departed completely from all previous designs. Out of our Ritz Carlton discussion came a tall, narrow, vertical antenna that scanned by being mechanically rocked and a horizontal antenna that scanned its pattern left and right. The beaver-tail beam from the vertical antenna would be so narrow that when its main lobe pointed at a plane very little energy would spill even one degree away, eliminating the possibility that the system would confuse the plane with its reflections. Alfred suggested switching a single radar transmitter between the two antennas four times a scan cycle. The three principles Alfred and I came up with on that memorable evening have been basic to GCA technology ever since. I was allowed to leave for home just before midnight.

The fact that Alfred Loomis had played a key role in its conception didn't hurt the new GCA design at the laboratory. After we had worked it out, Alfred dropped by to urge us to build a demonstration model as quickly as possible. If we did, he said, he would have ten prototypes built with Office of Scientific Research and Development (OSRD) funds so that the Army, the Navy, and the British could train with the prototypes while full-scale production got under way.

If someone wants government support today, he first has to write a lengthy proposal. Committees then judge his proposal in competition with similar documents from other sources. In the unlikely event that a proposal is accepted, it has to be worked into budgets that measure their time frames in years. On the basis of my experience in wartime, I'm sure we waste far greater sums of money on bureaucratic review, delay, and red tape than we could possibly lose if a project occasionally failed or a dishonest outfit defrauded. In the process we squander the enthusiasm of the originators, which may be an even greater loss.

The building of the Mark I GCA truck system was one of the happiest times in my life. We all believed strongly in what we were doing, and work blazed ahead. The laboratory had grown so rapidly that it expanded in the summer of 1942 into a modern concrete building, number 24, and a large wooden building, number 22. As a division head I was given an office in Building 24 directly above Lee DuBridge, in what was called Admiral's Alley. I never used my Admiral's Alley office. My real office commanded the center of the action in Building 22, which was only roughly finished and reminded me of the Berkeley Radiation Laboratory. I could drill a hole in the wall anytime I felt like it.

The Mark I GCA system used two trucks parked halfway down the runway, fifty feet from the left edge from the pilot's point of view. The larger of the trucks carried a gasoline-driven generator directly behind the driver's seat. Next came the azimuth antenna, which looked downwind at the approaching aircraft, rocking left and right. At the rear rose the tall elevation antenna, scanning up and down. The area search antenna was mounted above the generator. (The Air Force's experts insisted this antenna wasn't needed, but GCA would have been a dismal failure had we taken their advice.) The radar engineer monitored the antenna transmitters and the other electronic equipment from a narrow passageway along the side of the truck.

The GCA control truck, parked ahead of the antenna truck, looked in the direction of the landing aircraft. It housed the screens that displayed the radar signals and the controller's communications system. We ran the system for the first time as a complete unit in East Boston in late November 1942, and attempted on December 1 to land our first plane. We weren't successful—glitches in the equipment and too many observers on hand. We moved for five weeks to Quonset Point, where we were very successful, then back to East Boston. I did most of the controlling—talking to the pilots—and developed that technique by trial and error. I also had the pleasure of being the first civilian to make a low approach on GCA, under the hood, with Larry Johnston as my controller and Bruce Griffin as my check pilot.

The observers the Air Force and the RAF sent us apparently liked what they saw. On February 10 General Harold McClelland, the director of the Air Force Technical Services, asked us to drive the trucks to Washington for a demonstration on Valentine's Day.

The drive shook loose enough connections to postpone the test by twenty-four hours. A large group of high-ranking Army, Navy, and RAF officers showed up on the morning of the fifteenth. The high-voltage vacuum tubes in the transmitters kept blowing out for reasons we never understood, and I postponed the tests. The next morning the same group assembled, more problems developed, and I had to ask the military once again to come back the next day.

Most of our crew stayed up all night troubleshooting. Our standard crowd of high brass showed up gamely once again on the morning of the seventeenth. Larry Johnston had been a graduate student of mine at Berkeley; he had joined me at MIT and is listed on the GCA patent as coinventor. As I greeted the officers in a meeting room in the basement of the National Airport building, he whispered that the system was again out of operation but that I should entertain the crowd while he tried to get us back on the air again. He reasoned that if the brass left now we would never persuade them to return. As usual, the problem was burned-out vacuum tubes. The closest source of that type of tube happened to be the Anacostia Naval Air Station, directly across the Potomac. Our pilot took off from National, skimmed the river without even bothering to retract his landing gear, and landed at Anacostia. He was back within minutes.

While all this was happening, I did what I've seen television news anchors do many times since: I ad-libbed. After an uncomfortably long time Larry came in to say that the area search system was working but that the precision final-approach radars weren't up yet. He proposed that we show our visitors how we could vector an aircraft into a final glide path while we tried to fix the rest of the system.

With great relief I announced that we would now move out onto the field to demonstrate ground-controlled approach. By the time we had vectored the first plane into the glide path area, the final-approach transmitter was functioning. I took over as controller and gave the pilots landing instructions. Our observers listened by loudspeaker and watched the planes respond. We demonstrated that the aircraft really were under my control even though I could only follow them on radar. The high point of the demonstration was the approach of a colonel whom General McClelland had told only to get up into the air, tune his radio to a certain frequency, and do whatever he was told by some voice from the ground. The colonel had never heard of GCA but was an experienced instrument pilot. He first checked me out by changing his altitude and his heading. After each change I told him what he had done. He made several perfect approaches and then landed under the hood.

The demonstrations continued for a week, but the sales campaign succeeded that first day. The Army and Navy immediately ordered hundreds of units each. Fliers whose previous necessity in bad weather had been simply to bail out now had a good chance of landing. I've flown in and out of Washington National countless times since that February in 1943. Every time I've done so I've felt a chill remembering how close we came there to failing at the very brink of success.

When I returned to the laboratory after my summer of gallbladder troubles, my free-agent status had ended. I was made group leader of a potpourri known as Jamming, Beacons, and Blind Landing. I went on to invent not only GCA but also a microwave early-warning system, MEW, and a blind-bombing system, EAGLE. When the laboratory was divisionalized, I was made head of division 7, Special Systems, also known as Luie's Gadgets. All the other divisions had functional titles—Transmitter Components, Airborne Systems, and so on. Mine was organized to develop and build the systems I invented. (A definitive article on the MIT Radiation Laboratory in the November 1945 issue of *Fortune* included a laudatory section about MEW entitled "Alvarez's Folly"; that article and others made it clear that my work had an impact on the war.)

American bombers in World War II used the Norden bombsight. It was so secret that very few people even found out how it worked, and it was reputed to be so accurate it could aim bombs "into a pickle barrel" from thirty thousand feet. As part of my involvement in radar bombing, I requisitioned a Norden bombsight and a Norden trainer, a vehicle that carried the bombsight and bombardier eight feet off the floor of a hangar or a garage to simulate a bombing run. The target was a cross marked either on the floor or on a bug, which was a small remote-controlled box with wheels that could be programmed to simulate a moving target or crosswind effects. I learned the workings of the Norden bombsight in great detail because I was active in designing the EAGLE computer.

The Navy had originally purchased the Norden for use against large targets like battleships. To operate it, the bombardier clutched it in, activating a vertical needle on the instrument panel. The pilot steered his plane to keep the needle centered, thereby maintaining an accurate heading for several minutes. The bombardier twisted knobs until mechanically transported cross hairs remained fixed on the target. After that he simply waited for the bombsight to release the bombs automatically at the computed dropping point.

The Norden was extremely accurate in the games the Navy and Air Corps bombardiers played before the United States entered the war, one bombing group challenging another to an accuracy contest, dropping dummy bombs on a cross marked out on some convenient desert to win a few cases of beer. In wartime it was distinctly unhealthy to set an airplane on a fixed course and maintain that course for several minutes while the bombsight was being tweaked up. This straight path, with no jinking, made the bomber a perfect target for enemy antiaircraft batteries because the antiaircraft computers could predict where the bomber would be when the flak arrived. As the war progressed, bombing runs became shorter and the accuracy of the Norden bombsight correspondingly declined. The U.S. Strategic Bombing Survey found at the end of the war that the average target miss in Europe was nearly one mile. It would have been worse if the survey had measured the separation of impact and aiming points; instead it measured the separation of impact from the center of the nearest town, which might not have been the target.

The first three atomic bombings—Hiroshima, Nagasaki, and Bikini—further confirm these poor results. Hiroshima was hit almost within pickle-barrel range, but the bomb missed by two miles at Nagasaki. It missed by about a mile at Bikini despite the employment of one of the Air Force's best bombardiers and the absence of any enemy challenge at that 1946 demonstration.

In England the RAF bomb aimers managed a fair amount of joshing about the Norden bombsight. We Americans, they said, must build the world's largest pickle barrels. The impact-predicting RAF bombsight was much better suited to combat. It displayed a spot of light that appeared to move over the ground to show where bombs would land if they were dropped at that moment. The bomber's course toward the target could be erratic as long as the spot of light passed over the target at some time during the run—at which instant the bomb aimer could release his bombs. The bombsight measured the aircraft's acceleration and velocity relative to the fixed ground and included them in its computation.

I got to be a good bombardier with the Norden trainer. Those of us who were developing EAGLE practiced with it. We followed Navy tradition and converted it into a gambling device, making the target a fifty-cent piece forfeited to whoever came closest with his dummy bomb. I once visited Mr. Norden at his company headquarters in Brooklyn. He was an elderly Dutchman who had grown up in Java and who remembered the explosion of the Krakatoa volcano in 1883. He had personally designed every part of the bombsight, and on a drafting board on his very large desk in his enormous, paneled office he was designing its latest modification when I visited him. I was amazed to notice that he was drawing in the diagonal lines on the screws, a practice our draftsmen had abandoned years earlier. His factory contained more precision machinery than I had ever seen; it was a good example of what happens to military procurement when money is no object.

EAGLE became operational only at the end of the war, but a mission of fifty-nine EAGLE-equipped B–29's based at Tinian then practically obliterated a Japanese target. (I had a good visit with the EAGLE bombardiers just after our 509th Group put them out of business.) The subsequent commendation noted that the mission achieved 95 percent destruction, the most successful radar bombing in the command. Postwar analysis showed that planes equipped with EAGLE could bomb at night and through overcast with greater accuracy than Norden bombsights could deliver in broad daylight.

To demonstrate our ground-controlled approach system to the British, I got another chance to go to England. So many of our people would eventually be working there full-time that we later opened what was called the British Branch of the Radiation Laboratory at Great Malvern, some fifty miles west of London, where the British radar laboratory was also located. Since I would represent the entire laboratory, I spent several weeks finding out in detail what all our groups were working on and what progress they had made.

I left for England on June 10, four years after the first commercial transatlantic flight. It was still an adventure to fly the Atlantic on a Pan American Clipper. About twenty-five passengers assembled in the lounge at LaGuardia Airport and were delivered to the Clipper a little after midnight. The Boeing aircraft was luxurious, the largest I had ever flown. We taxied out into Long Island Sound for a long run on the water. With a cruising speed of 120 miles per hour we made Nova Scotia in time for breakfast. Like the Concorde, on which I later flew, the Clipper was something of an exclusive club. Among the fellow passengers I met that morning were Hugh Baillie of United Press and his boss Roy Howard, who had gained notoriety in the First World War by announcing the German surrender three days before it took place; MacKinlay Kantor, the novelist and correspondent; a Hollywood producer; and the advertising executive William Benton, later to become a U.S. senator and the publisher of the *Encyclopaedia Britannica.* We had the University of Chicago in common—I an alumnus, he a trustee.

Midmorning we headed for Botwood, Newfoundland, from which we took off again in early evening following the great-circle route pioneered by Charles Lindbergh. The crossing to Ireland was turbulent; many of the passengers were sick. We flew below the fog banks, a few hundred feet above the waves. Shortly after sunrise I spotted Ireland, the greenest landscape I had ever seen.

The final leg of my crossing was by RAF Sunderland flying boat, an aircraft almost as large as the Clipper but accommodated like the Spartan military transport it was. That was the first time I had flown aboard a military plane in a war zone. We landed that evening at Poole, were driven to Southampton, and took the boat train to London, forty hours out of New York. In Hyde Park at ten o'clock at night I noticed American GIs playing softball in broad daylight.

Barrage balloons tethered on thousand-foot cables guarded the city. Estimating a typical bomber wingspan and the number of balloons per square mile, I calculated that the chance of a bomber's crossing London undamaged below one thousand feet was only about 50 percent. If the bomber flew above the balloons, the radar-controlled antiaircraft guns had a good chance of shooting it down. The Battle of Britain was over, but the Germans still frequently sent raids. I could turn off my hotel room light, open the heavy blackout drapes, and watch German bombers caught in the British searchlights. I had seen antiaircraft fire only in black-and-white newsreels, so I was unprepared for the red flash when the shells burst. I had become an Anglophile when I helped found the MIT Radiation Laboratory and discovered the extraordinary work our British friends had been doing. My experiences in England increased my admiration even more.

Meals, however, were a trial. The choice at breakfast at my London hotel was sausage or eggs. The sausage, I was assured, consisted by law of at least 50 percent meat. Judging by texture and taste, I took the other 50 percent to be sawdust. The eggs were powdered and almost inedible. Toast always arrived cold in a silver rack, which I thought of as a toast cooler. The coffee was bad, though not as bad as the ersatz product the Germans were drinking. I'm not criticizing, only complaining. War is hell.

The NDRC maintained a small London office near my hotel. I spent my first weeks working there and visiting military establishments. My first official duty was attending a meeting at the Admiralty, where the most junior officer was a rear admiral. The subject under discussion was Alfred Loomis's pulsed navigation system, Loran. To my surprise I knew more about it than anyone in the room and was called upon to explain it in detail to an obviously interested audience.

My GCA group would be stationed at Elsham Wolds, an RAF aerodrome, later in the summer. A British Pathfinder pilot delivered me there in a new B–17 Flying Fortress that a U.S. Air Force friend had lent him for evaluation—a Pathfinder was an elite pilot who flew ahead of RAF bomber groups to mark bombing targets with parachute flares. I rode copilot. The pilot liked the new aircraft very much, exploring the knobs and switches and pronouncing it a "wizard kite." The bomb bay puzzled him, however. It was about right for a forward bomb bay, he thought, but he could locate no counterpart in the rear. He found it hard to believe that an aircraft the size of the B–17 had been designed to carry such a small volume of bombs. The British bombers, considerably larger than ours, carried enormous loads of bombs, many of them primitively unfinned and unstreamlined. The British similarly bypassed such amenities as copilots and provided only one set of flight controls. A wounded pilot was simply removed from his seat and replaced by a crew member, whose flight training might well be minimal.

The station commander at Elsham Wolds was happy to allow GCA tests on his base so long as we didn't interfere with his bombing missions. During the first week of July I met my GCA colleagues at Victoria Station, just as I had been met before, and introduced them to wartime England. We located our GCA set, picked up RAF drivers to deliver the two trucks, and met our WAAF contingent, the women who would be our drivers and baby-sitters for the next two months. Our WAAFs were all corporals, and one was a Lady. They shuttled us back and forth each day from the Angel Hotel, our comfortable quarters in Brigg, to Elsham Wolds. The hotel proprietress, a very special person, kept her own hens; we quickly became friends. She provided each of us with the unbelievable treat of a fresh egg for breakfast every morning. I remember as the high point of an RAF party the auctioning off of a pair of fresh eggs; the station commander successfully bid five pounds, and everyone thought he'd done well.

For six weeks I lived in intimate association with men whose daily job was fighting a war and who too frequently disappeared from the dining room, never to be seen again. When the bombers were flying in bad weather, we stood by. In good weather we repaired to the local pub and practiced tolerating British beer. We had brought along a large supply of spare vacuum tubes, and our GCA tests were successful beyond expectation. We landed every type of plane the RAF owned, every rank of pilot, from sergeant to air chief marshall, and on several occasions the entire bomber squadron returning from a mission over Germany. The mission landings were important tests. It's one thing to approach a foggy airport on instruments after a good night's sleep, quite another to do so after eight hours over hostile territory in a damaged aircraft with wounded aboard. We brought them in, those who came back.

When Air Chief Marshall Sir Ludlow-Hewitt visited Elsham Wolds, he was impressed with the great variety of planes and pilots we landed but noticed that we had never bagged a tired fighter pilot. He proceeded to order one up, a Typhoon returning from long hours patrolling the North Sea. Squadron Leader N. P. Mingard, whom I had trained, did the controlling; I had the pleasure of standing outside the

GCA truck as the air chief marshall listened in. The Typhoon pilot was a free Pole who had escaped Poland and joined the RAF. He didn't know much English, but enough. He appeared out of the overcast at two hundred feet directly over the runway center line, made a touch and go, and went around for a second try. As he touched down the second time, Mingard asked him what he thought of the system. "Wizard, absolutely wizard!" he shouted and proceeded to show his pleasure by firing his two forty-millimeter cannons down the runway. No one had ever fired such large-caliber ammunition only a few feet in front of my eyes, and I was impressed.

The normal complement at Elsham Wolds was eighteen Lancasters. Mission losses dictated that fewer were usually on hand. The Lanc was a vast, black, four-engined bomber with twin vertical stabilizers and a huge bomb bay. The standard RAF tour of duty was sixty-five missions, which made the odds for survival extremely small. The crews were understandably fatalistic.

Standing at the officers' club bar one night, I was surprised by a powerful explosion that shook the building and rattled the windows. We ran outside to find a thick black plume rising from the pad where Lancaster C-Charlie was normally parked. A two-ton cookie, one of the unfinned cans of high explosive that the RAF dropped on German industry, had gone off. Two of the Lanc's large Rolls-Royce engines had been blown several hundred yards away. A third turned up the next day half a mile outside the aerodrome. The fourth was still missing when I left England. Not one of C-Charlie's crewmen was injured—they knew their cookie was on fire and beat a hasty retreat —but a small piece of flying metal penetrated a nearby Lanc that was trying to taxi out of range and killed the radio operator. A line drawn through C-Charlie's name on the operation's room status chart would have written it off, but in the ops room the next day it was listed with hard-earned British understatement as merely US, unserviceable. Everyone in housekeeping turned up with accumulations of broken glasses, crockery, and window panes saved for just such an occasion. The RAF inspector gave them full credit from the RAF insurance fund.

When I toured the British laboratories and talked to their staffs, I was struck by the difference in attitude and atmosphere from MIT. The British were grimly aware of the seriousness of the war. I particularly remember the propeller of a German bomber that was mounted on the wall of the office that corresponded to Louie Ridenour's 584 office at MIT. They had personal experience with injury and death.

Many of their counterparts in the United States were being well paid to work with the best equipment they had ever seen doing very nearly what they had always wanted to do. The war was far away, and most of them had little or no contact with anyone who had seen a war zone.

Near the end of my tour I was invited to attend a briefing for an RAF night-bombing mission. Squadron Leader Mingard and I took off in the morning for a Bomber Command base near Cambridge, flying a Miles Master. Cruising at five thousand feet, we heard a grinding crash from up front, followed by billowing black smoke. The smoke poured through Mingard's cockpit and began seeping into mine. The propeller had stopped rotating, and we were abruptly flying a glider with the aerodynamic properties of a brick. In the sudden and surprising silence, I could talk directly to Mingard without using the gasport —the speaking tube. I asked him what had gone wrong. The engine has packed up, he answered, and we're going downstairs. He then flew a series of steep, alternating sideslips to blow the black smoke away so that he could see. He surprisingly found no airstrip but located an open field.

Mingard managed a nice approach at just the right altitude over the fence, but closer to the field we saw that it was crossed with parallel ridges running at right angles to our approach. Our left wheel struck a ridge and collapsed. That dropped the left wingtip, which hit the ground and tore away. The right wing dropped in turn, the right strut collapsed, and part of the wing ripped off. We came very quickly to a full stop, wings strewn behind us.

I was wearing my Sutton harness, which attached by a spring behind me to the fuselage, and wasn't even scratched. Mingard hadn't bothered to strap on his harness and was bleeding from a banged forehead. The wood-and-canvas plane fortunately didn't catch fire. (It's not an accident that both of the planes I have owned have had two engines.) Within minutes of our crash a swarm of small boys came running across the field to see what they believed to be a plane shot down by the Germans. They were more than a little surprised to see a civilian climb out. We got a ride to the nearest town, called for rescue, and flew on. Mingard announced afterward that if the smoke had been 10 percent worse we would have had to bail out; he couldn't have found his way to land. (That's how closely I missed my second parachute jump.)

At the Bomber Command base we were hurried to the briefing room just in time to hear the intelligence officer announce that the target for the night was Hamburg. The raid, which churned up a fire storm, would be the largest yet mounted by any air force, 791 planes that July night, 787 following up two nights later. WINDOW—tons of strips of aluminum foil—would be used for the first time to jam German radar, and the radar countermeasures officer briefed the crews on the use of this previously secret invention. The bombers had not yet been fitted with dispensers; crewmen at regular intervals simply kicked the loosened bales of foil out the side door.

We watched the bombers take off, ate dinner, and settled in to wait. It was fascinating to sit in on the debriefing when the air crews returned. Intelligence officers fluent in German had eavesdropped on the transmissions between the German GCI radar operators and the night fighters they directed. WINDOW had effectively blinded the radars. The British lost only twelve aircraft that night, welcome news to those who had worked hard to develop the jamming material and for months had urged its operational use. Unfortunately, three of those twelve aircraft came from the Elsham Wolds squadron, the first losses it had sustained in more than a week; the squadron commander thought WINDOW an abomination and let his crews know they were free to forget to kick the bales overboard. The attribution of significance to a statistical fluctuation like this is one of the hazards of a scientific career, and many otherwise competent scientists who carelessly accept such evidence have had their reputations ruined.

I visited a Chain-Home station on the cliffs of Dover; I visited the Pathfinders; I visited a U.S. air base, where I discovered myself sufficiently Anglicized to be offended by chest loads of ribbons on novice crews and raucous talk in the officers' mess. I had read everything Ernest Rutherford had ever written, much of it several times over, so a visit to the Cavendish Laboratory was a must. There I met an old friend, Bernard Kinsey, who was using the Cavendish cyclotron to measure neutron cross sections for what the British called Tube Alloys, their atomic-bomb project. Kinsey explained in great detail what he was doing without even bothering to ask me if I was cleared to hear such secrets. I learned thereby that the British were unimpressed with the elaborate rules the U.S. security people had invented to prevent people who did not have a "need to know" from learning something from their friends. Those rules were mostly an annoyance, and almost everyone I worked with paid little attention to them within the community of scientists. To my knowledge such defiant behavior did not compromise our security; to the contrary, it enhanced it by getting the job done. Klaus Fuchs, the Soviet spy, was cleared for access to the American Manhattan Project as part of a British delegation sent over late in 1943, but the British inexplicably ignored his affiliation with the German Communist party.

We finished our Elsham Wolds tests the last week in August, and the RAF arranged to transfer the GCA unit to its Coastal Command, where it would help land bombers flying over the Atlantic on antisubmarine patrol. I passed my leadership along to George Comstock, who in turn passed it to a young RAF officer at Land's End airfield named Arthur Clarke. Arthur told the story of that particular GCA set after it became British property in his novel *Glide Path,* which he dedicated to George and me. Arthur predicted a physics Nobel Prize by 1958 for the character in the novel he modeled after me; when I finally received mine, in 1968, the author of *Childhood's End* and *2001,* by now an old friend, sent along congratulations and reminded me of his prescience. (Arthur was the first person to suggest the use for communication of satellites positioned at such an altitude that they orbit directly over a particular spot on the equator—geostationary satellites. I had the pleasure of nominating him successfully for the Marconi Award for this achievement. Communications engineers now refer to the location of the several hundred geostationary satellites as the "Clarke Ring.")

On my final day in London the Air Ministry invited all the officers and civilians I had worked with during the past three months to a reception in my honor. I took that fine send-off as a measure of a job the RAF thought well done. It was the last I would see most of my GCA comrades for a long time to come. We did have a wonderful thirtieth-anniversary celebration in Boston in 1971; the album of photographs the group gave me was signed by twenty-five friends who made GCA possible. Arthur Clarke traveled halfway around the world from his home in Sri Lanka to attend.

SIX

Working with Fermi

THROUGH THE YEARS of radar work I sometimes participated behind the scenes in the developing American effort to make an atomic bomb. Eventually I joined that project full-time.

Ernest Lawrence visited MIT frequently in the early days of the Radiation Laboratory and kept me informed. He told me about Glenn Seaborg's discovery of plutonium and Seaborg and Emilio Segrè's finding that the new transuranium element fissioned even more readily than uranium. I knew about the conversion of the 37-inch cyclotron into a mass spectrometer that separated small amounts of uranium 235 from more plentiful U-238 so that the relative capacity of the two isotopes for fission and for neutron capture could be measured. As this work developed, Ernest realized that useful quantities of U-235 could be separated by a sufficient number of large-scale mass spectrometers and began championing that heroic approach to isotope separation.

Robert Oppenheimer invited me to attend a meeting at Columbia University in the fall of 1941. There I saw one of the exponential graphite and uranium oxide piles—subcritical nuclear reactors—that Enrico Fermi, Leo Szilard, Herb Anderson, and Walter Zinn were assembling in the basement of Schermerhorn Hall. I had no clearance for the uranium project and wasn't even supposed to know that it existed, but I sat in on detailed technical discussions at Columbia that day.

I was involved again several weeks after Pearl Harbor. At a meeting in Washington the day before the Japanese attack, the OSRD had approved a major expansion of the bomb program. The development of methods to separate uranium isotopes and to breed plutonium in quantity was the necessary next phase. Ernest Lawrence would head the mass-spectrometer program. Harold Urey at Columbia would pursue isotope separation by gaseous diffusion. Arthur Compton was responsible for plutonium production.

Arthur phoned me in Cambridge and asked me if I would come to Chicago to talk with him about a matter of great urgency. As soon as I could arrange air travel priority, I was there. I stayed in Arthur's home for several days, sleeping in his young son John's room on the third floor. We discussed intensely the problems of getting a wartime research project under way—how it should be organized and what sort of people it needed. As a researcher, Arthur had been a lone wolf. He knew I had experience in team research and felt that as one of his old pupils I could give him good advice.

The major decision he needed to make immediately was what to do with Fermi's group at Columbia. There was concern at that early point in the war that our coastal cities might be attacked; it was also hard to see how Columbia could stretch its resources to support two large uranium-related projects, Urey's and Fermi's both. Robert Maynard Hutchins, the president of the University of Chicago, had promised Arthur that the university would fully support the Columbia group if it moved there. Arthur and I discussed such a move at great length. I argued against it. Fermi's group seemed to be doing well at Columbia. I felt a move would set them back by several months. Moreover, Chicago did not have any tradition of work in this field of physics.

In the course of my visit Ernest Lawrence came to Chicago to try to persuade Arthur to move the pile program to Berkeley. His argument was convincing: his Radiation Laboratory had support facilities for such a project and experience in the administration of large projects.

Ernest and Arthur were scientific peers. I fitted in several rungs below them and felt a strong emotional attachment to each. I remember in particular my surprise at their behavior during a very long and decisive meeting in Arthur's bedroom—he had the flu and felt miserable. Ernest argued in favor of moving everyone to Berkeley. Arthur resisted, sensing that in Berkeley he would be reduced to a cog in Ernest's large machine. In all the years I had been Arthur's student, I had never seen him fight so hard for anything. Ernest had a legitimate interest in the pile program. The work at Columbia had been begun to develop nuclear power with no application to the war. The pile's potential to breed a fissile element for bomb use that could be separated chemically from natural uranium, a far more straightforward process than isotope separation, became apparent because of the Berkeley plutonium work. (On the other hand, I never sensed that Ernest considered the plutonium project to be competitive with his mass-spectrometer project, although it did in fact compete intensely for manpower and materials.)

The discussion between the two men became increasingly acrimonious. Abruptly Ernest declared that if everything wasn't moved to Berkeley he would bet a thousand dollars that a chain reaction wouldn't be achieved within the year. I was shocked; I'd never heard either Ernest or Arthur bet so much as a nickel on anything. Arthur thought for a minute and announced that he would take the bet. Then they both looked at me, obviously embarrassed that I had heard these intemperate remarks and witnessed such uncharacteristic behavior. When they cooled down, they canceled the bet. That was unfortunate for Arthur. He moved Fermi to Chicago, Fermi and his group assembled the first uranium-graphite pile in a squash court under the west stands of Stagg Field, and it became the first operating nuclear reactor on December 2, 1942. He would have won the bet with three weeks to spare.

It was Ernest's idea to build a thousand enormous mass spectrometers, dubbed calutrons, to separate U-235 from U-238. The scale of his proposal led everyone to question his sanity at first, but with a series of increasingly large test devices he demonstrated that he was on the right track. Understanding that the 184-inch cyclotron could be converted to a large-scale calutron, everyone pitched in to complete construction ahead of schedule. The big magnet went up first and its building around it. When I visited the new laboratory on the hill in March 1943, before I went to England, I found all this activity amazing. My old love for magnets, vacuum plumbing, and cyclotron electronics was rekindled. I quickly convinced myself that I had contributed all I could to radar and that my MIT colleagues could carry on perfectly well without me. That was a reasonably accurate assessment; in any case, the organization of the MIT laboratory had evolved to the point where I would have found it impossible to start any new projects. The lab had separate divisions for land, sea, and airborne radar systems; the only region left for invention, I told myself, was underground. I had been a chief long enough to know that I didn't want to spend the rest of my life as one. The possibility of returning to Berkeley as the active Indian I once had been was exciting.

Ernest had a difficult task in the putting together of an effective team to do the big job he had in mind, especially since he'd sent so many of his trusted colleagues and former students off to MIT. He did find competent people to fill the key roles. It was satisfying to stand behind a control table and watch them operate the calutrons, tuning the meters to get maximum separation of the U-235 and U-238 ion beams (the mixed ion beam curved through a magnetic field, where the slight difference in the masses of the two isotopes caused the ions of each to follow slightly different paths, from which they could be separately collected). I wanted to sit in those chairs and do that work.

Great salesman that he was, Ernest encouraged my interest. He wanted me back on his team and hinted enough about the challenges of the separation project to persuade me to decide to leave radar. At MIT I told people of my decision. Bob Bacher and Ken Bainbridge had just signed up to go to Los Alamos, New Mexico, where Robert Oppenheimer was establishing a new secret laboratory to design and build the first atomic bombs. In the course of a long talk Bob and Ken convinced me that I should join them; by the time I got back from England, it appeared, the calutron project would have advanced beyond the development stage into industrial production at Oak Ridge, Tennessee.

They arranged for me to meet with Robert Oppenheimer in New York. Robert and I spent several hours considering how I could help. He was also a fantastic salesman and had no difficulty persuading me to come to New Mexico; I was already sold on the idea. Remarkably, in retrospect, he left me with the impression that the problem of making an atomic bomb was essentially solved and that the major interest at Los Alamos was the thermonuclear—the Super, as it was always called—the hydrogen bomb. Much of our conversation was devoted to the challenging problems that would have to be overcome in pursuit of the H-bomb. Discussing them excited Robert, and he gave me more information than I could assimilate. (In the spring of

1943 only the problems of making the U-235 bomb were well in hand. But no one could have anticipated how difficult it would turn out to be to design a detonation system for plutonium 239; not enough plutonium was yet available to make the necessary measurements.)

At the end of a long and very pleasant conversation with Robert, I agreed to join his laboratory when I returned from England. I wrote Ernest of my decision, which I'm sure he agreed was the correct one. Lee DuBridge was shocked when he learned that three of his division heads were leaving for what by then was known as the Manhattan Project.

While I was in England, Robert cabled to ask if I would work with Enrico Fermi in Chicago for a time rather than proceeding directly to New Mexico. Fermi had wanted Emilio Segrè, his former student and collaborator, to leave Los Alamos and join him at the newly formed Argonne Laboratory. Robert preferred to trade a bird in the bush for a bird in the hand. I was overjoyed at the prospect of working with Fermi, for whom I had enormous admiration. I was flattered that he would consider me an acceptable substitute for his trusted colleague and longtime associate.

When I reported for duty in Chicago, I found the two physics buildings where I had spent so many happy years thoroughly secured. They headquartered the Metallurgical Laboratory, a division of the Manhattan District of the U.S. Army Corps of Engineers, which had assumed responsibility for the rapidly expanding bomb program. Arthur Compton directed the Met Lab; in late 1943 he was spending most of his time at Oak Ridge, where an intermediate-scale air-cooled reactor was under construction. It was larger and much more powerful than the first pile tested in the Stagg Field west stands had been. Chicago Pile No. 1 (CP–1) had been dismantled and moved to a new laboratory in the Argonne National Forest twenty miles southwest of Chicago. CP–1 had no cooling or shielding, so it could be run only at power levels of a watt or less. At Argonne it had been reassembled as CP–2 inside concrete shielding with forced-air cooling supplied by two very large fans that blew air through the interstices of its graphite bricks. With shielding and cooling it could operate at higher power.

I checked into the Met Lab personnel office and found I'd be spending most of my time at Argonne, where Fermi was working. A bus, the Blue Flash, carried me back and forth one hour each way almost every day for the next six months, but my first trip out was with Enrico Fermi and his Army-assigned bodyguard John Baudino, a young Italian-American lawyer who added notably to Enrico's store of baseball lore and American slang.

CP–2 was a thirty-foot cube of graphite bricks set with slugs of uranium oxide and metal surrounded by a thick concrete shield. Enrico led the way to the top of the reactor to show me his thermal column, an opening into the system from which emerged a flux of pure thermal neutrons. Unlike my prewar time-of-flight arrangement, Fermi's thermal column was not only operationally pure; the neutrons that emerged had diffused through enough graphite to have left their faster ancestors far behind. With it experiments could be managed in minutes that had previously taken most of a year.

The reactor control room was similar to a cyclotron's but simpler. The moving of control rods made of neutron-absorbing cadmium in and out of the pile decreased or increased the rate of reaction by decreasing or increasing the flux of neutrons available for fission. The period of increase could easily be changed from minutes to hours to infinity. At infinity the neutron intensity was remarkably constant. I found it easy to learn to run the pile, which was boring, I thought, compared with running a cyclotron. The system was extremely sensitive to changes in atmospheric pressure, which changed the content of neutron-absorbing nitrogen in the small spaces between the graphite bricks. I once spent several days trying to invent a barometer sensitive enough to anticipate such changes and adjust the control rods accordingly. I concluded that the reactor itself was by far the world's most sensitive barometer and gave up.

The reactor operators at Argonne were testing the graphite bars and uranium slugs that would go into the Oak Ridge and Hanford, Washington, reactors. Boron and other impurities with large appetites for neutrons could poison the chain reaction. The troubles they caused were proportional to their impurity levels times their neutron-capture probabilities, so when the latter were large, chemical assays of impurity levels weren't sufficiently sensitive. The pile could measure poisoning directly. The operators substituted newly manufactured bars for standard CP–2 bars and found the control rod positions that kept the neutron level constant. A comparison of these test positions with the standard positions immediately measured bar impurities. Uranium slugs were tested similarly. It was dull work but an essential service to the cause.

Enrico Fermi was an amazing man, quite unlike any other physicist I have ever met. His lectures, as I have mentioned, were perfectly organized and almost entirely mathematical. At Argonne he usually worked by himself in his own room, only occasionally inviting us in to share his thoughts or to ask our advice, which he seldom needed. I did have the pleasure of lunching with him almost daily for six months.

One of the few times I was able to help him began over lunch one day in the small cafeteria in the reactor building. Herb Anderson was there, as were Fermi's protégé Leona Marshall and her husband, John. Enrico said that in solids we should be able to reflect neutrons as we do X-rays, at grazing incidence. That neutrons might have an index of refraction was a new idea to us, and we hadn't considered the possibility of reflecting them. After lunch Enrico invited us into his workroom. When we were settled around his desk, he remarked that the equation for the refractive index of neutrons should be similar to that for X-rays, except that the classical radius of the electron would be replaced by the square root of the neutron's cross section for scattering (a measure of the probability that scattering will occur). He said that the equation for the refractive index was 1 minus something small, but he couldn't remember the exact formula.

I piped up that it was in Arthur Compton and Samuel Allison's standard text, which I knew was in the next room, and that I'd go and get it. Enrico said not to bother; he'd derive it.

As a student of Compton's, I had thought long and deeply about X-rays, but I had never seen the refractive-index formula derived from basic principles. Enrico wrote James Clerk Maxwell's classic electromagnetic field equations on the blackboard and then in about six separate steps derived the formula. The most remarkable aspect of this tour de force was that Enrico worked through his derivation line by line at a constant rate, as if he were copying it out of a book. That night at home I reproduced it and was quite pleased with myself. If one step was easy enough to allow me to go faster than he did, the next was so difficult that I could never have managed it alone. But Enrico worked the difficult steps at exactly the same rate he worked the easy ones.

Having derived the refractive-index formula, Enrico plugged numbers into it and then, using the six-inch slide rule that was his constant companion, estimated that the difference between 1 and the index of refraction was about one-millionth, just as I knew it was for X-rays in ordinary materials. He frowned. Too bad, he said, we certainly wouldn't be able to measure angles of reflection of neutron beams that were one-millionth of a radian.

I thought a moment and commented that the same factor applied to X-rays and that, since their reflection was so easy to observe, the reflection angle must go like the square root of the decrement. Enrico immediately perked up and said I was right. It was the cosine of the grazing angle that was involved and not the sine. Since Enrico read little of the physics literature, he might not have realized that the decrement for X-rays was also one-millionth and that X-ray spectrometers couldn't measure such small angles. (After the war, he and Leona Marshall in a beautiful set of experiments using the Argonne heavy-water reactor determined the index of refraction of neutrons for many materials and showed how one could easily make neutron beams that were 100 percent polarized; Felix Bloch and I had managed after two months of hard work to achieve a polarization of only a few percent.)

Enrico's reference library consisted of only some half a dozen books, including his own works on thermodynamics and atomic spectroscopy, supplemented by a vast collection of elaborately cross-referenced notebooks containing formulas he had derived. When he read the title of an interesting theoretical paper in the *Physical Review,* he told me once, he tried to work out the problem first and then checked it against the author's results as stated in the abstract. Two other books on his shelf were the *Handbook of Chemistry and Physics,* which contained much that he could use, and the Jahnke-Emde *Tables of Functions,* which tabulated values of the more exotic functions that appear as solutions to differential equations. He was the only completely self-sufficient physicist I've ever met. Every other successful physicist I've known required give-and-take between himself and students or colleagues. Enrico got along perfectly well alone and gave of himself essentially to help his students and colleagues. In social situations he was retiring and didn't volunteer opinions; he seemed modest and self-effacing. But I'm sure that if I had asked him who in his opinion was the best physicist alive he would have thought for perhaps half a minute and nominated himself. He knew exactly where he fit into the world of physics, and he was honest enough to give the right answer.

Argonne served me valuably as a decompression chamber. I had gone to MIT as a young, hardworking experimental physicist with no administrative responsibilities. There I quickly became a group leader and division head responsible for the work of hundreds of scientists. I spent a lot of time dealing with the military brass. Those duties led me far away from the world of the laboratory scientist I had been. Without decompressing in the pure research environment at Argonne, I might have remained a full-time administrator, as so many of my wartime colleagues did.

I began by working alone, studying reactor theory and the advances physics had made while I was off developing radar. I had the great privilege then of learning Fermi's way of doing physics, which was entirely new to me.

Within a few weeks of my arrival Enrico suggested that I measure the number of neutrons produced from a standard radium-beryllium source. The number was already known to within a few percent. Enrico thought I could do considerably better than 1 percent by using the reactor as an intense source of neutrons. The experiment isn't important enough to describe in detail, but it had much to do with converting me from a chief back into an Indian. One piece of equipment I used was a large tank made of galvanized iron sheets soldered edge to edge. To keep the thin iron sheets from bulging when it was filled, the tank was reinforced with an external plywood frame; for my experiment it held borax and a manganese salt dissolved in water.

After much thought I soldered a spigot into the tank to make it possible to extract samples of the solution. I did a bad job of soldering, and the seam leaked. Since I was working alone and had no influence with the shop technicians, I had to fix the leak myself. I thought I could repair the leak with an acetylene torch and soft solder. Whenever I tried to heat the joint, though, the water boiled on the other side of the iron. That kept the joint at 212 degrees Fahrenheit, not hot enough to make the solder flow. I refused to give in to this basic demonstration of heat transfer and doggedly worked on. One of the shop technicians happened by and told me I could never repair the joint with the tank filled. Stubbornly I kept trying, and finally, after two frustrating days, I succeeded. Despite the frustration it was a wonderful experience. It convinced me that I wanted to be a technical man, hands-on, not the administrator I had become who would have called up a technician because he was too busy to bother with such petty details. Fixing that leak freed me to return to productive, active science again.

Brigadier General Leslie R. Groves was the Army engineer responsible for the entire Manhattan Project. Almost everyone disliked him heartily except the men who worked closely with him. Both Robert Oppenheimer and Ernest Lawrence told me how much they respected

General Groves. If the project had been a disaster, he would have shouldered much of the blame. Since it was a great success, he deserves, I believe, much of the credit.

I was so far down the Met Lab structure that I didn't appear on any organization chart; nevertheless, the general knew who I was. I was summoned one day to meet him in the Army district manager's office. He called me in eventually and asked me how we could find out whether the Germans were operating nuclear reactors. If they were and if we could locate them, then the Air Force could destroy them or at least interrupt their operation. General Groves told me to look into the problem and to report directly to him in a week. He cautioned me to let no one else in the laboratory know what I was doing.

I gave the general's problem much thought. The scheme I devised involved detecting the radioactive gases that reactors emit in normal operation. We could most easily detect xenon 133, I decided, which has a five-day half-life, during which it produces distinctive gamma and beta radiation. Xenon is a noble gas, and since its boiling point is much above those of nitrogen and of oxygen, it should be possible to build a scrubber that could process thousands of cubic meters of air and trap any radioactive xenon atoms that came along. The scrubber could be built into the front of an airplane and flown over Germany to sample the air mass. When the plane returned, the xenon atoms could be identified by their characteristic radioactivity. I believe I was the first to look into the monitoring of fission processes by flying sample-collecting aircraft, a form of intelligence gathering that became extremely important after the war.

I moved my office to Eckart Hall on the University of Chicago campus, where I had access to the fine physics library that had earlier had such an important impact on my career. With my Manhattan District triple-A priority, I could go anywhere I wanted and arrange to have anything I needed built. Since I needed to know more about the handling of rare gases, I traveled to General Electric in Cleveland, the country's largest processor of air because it used argon in its light bulbs. With the help of GE's rare-gas experts I designed equipment to fly in the bombardier's compartment of a Douglas A–26 and contracted with GE to build it. The system passed an air sample through activated charcoal. That trapped any xenon and radon but not nitrogen and oxygen. Radon, also a noble gas, is present in the atmosphere at a lower concentration than xenon and also has a much higher boiling point (as its name implies, it's a decay product of radium, which is in turn a decay product of uranium, which occurs naturally throughout the world in low concentration but great volume). After an overflight, we could heat the activated charcoal to boil off both the radon and the xenon into a stream of helium gas. Passing that gas stream again through activated charcoal at ice temperature would freeze out the radon but allow the xenon to get through. More activated charcoal, this time at dry-ice temperature, would absorb the xenon. With the helium pumped out, this last filter would be heated to drive off pure xenon. The resulting highly concentrated sample could then be examined for radioactive xenon, the presence of which would indicate an operating German nuclear reactor. The equipment did fly over Germany but detected no radioactive xenon, because, as we discovered near the end of the war, German scientists did not succeed in building a reactor large enough to initiate a chain reaction.

While I was developing this detection system, I got regular calls instructing me to meet General Groves at the railroad station in Chicago, where he paused for a few hours at a time on his regular transcontinental trips. The general held court in the waiting room. People summoned to meet him would approach him one by one, talk quietly for a few minutes, and then give way to the next person in line. I rather liked the general, whose methods were similar to those of the military officers I had dealt with for the past three years. He keenly appreciated the problems involved in trying to monitor German fission research, and once the project moved out of my hands into its development phase he always appeared to know more than I did about its progress.

During one of my meetings with General Groves, he informed me that he had appointed me to a secret group of radiation experts that would be flown immediately anyplace where the Germans or the Japanese might have dropped an atomic bomb. Apparently a long-range aircraft equipped with instruments for measuring radiation was standing by somewhere.

Besides Fermi I worked at Argonne with Herb Anderson and with Leona and John Marshall. I saw less of Wally Zinn; he was busy building the first nuclear reactor moderated with heavy water. The heavy water, made at Trail, British Columbia, was arriving in ever-larger containers. Wally worried that firemen might use their hoses if the building ever caught fire and spoil his precious tons of heavy water. He was prepared to watch the building burn down rather than risk letting the firemen in.

Those of us around Fermi had the pleasure of doing research that had little connection with the war. The rest of the Met Lab worked hard advancing the science and technology of nuclear reactors, with the goal of producing plutonium for nuclear weapons. My only contact with this main effort was a weekly information meeting in Eckart Commons that Sam Allison usually chaired. We heard reports on the progress of the Oak Ridge intermediate reactor and the big production reactors at Hanford. I was surprised to learn that such good physicists as Sam Allison and the theoretician Eugene Wigner spent most of their time poring over the Hanford blueprints. When I was in charge of large engineering projects, I made it a point not to examine blueprints in order to second-guess my engineers. I discussed a project at great length so that they came to appreciate the problems that I felt had to be solved, but I left them on their own to solve those problems. I operated on the belief that if engineers know that physicists are going to check their blueprints they won't be nearly so careful. I treated engineers as full partners with the responsibility for flawless design. These methods worked for me, and no serious engineering mistakes ever slowed down any of my projects.

In my last two weeks at Argonne I discovered the long-range alpha particles that accompany fission—the only real physics I did in five busy war years. I'm still amazed I didn't discover them the first time I looked, in 1938.

I enjoyed my stay in Chicago, but I was also embarrassed to be working so far from the war. When Bob Bacher showed up one day and asked me to join him at Los Alamos, I was relieved. I worked night and day on my long-range alpha experiment, and then I caught the Super Chief to Santa Fe.

SEVEN

Los Alamos and the Atomic Bomb

BOB BACHER recruited me to work at Los Alamos with the Harvard chemist and explosives expert George Kistiakowsky. The secret New Mexican laboratory was developing two methods for detonating nuclear explosives. One, which involved firing one piece of nuclear material against another inside the barrel of a cannon—the gun method—was progressing well. The other, squeezing a hollow piece of nuclear material to solid density by means of high explosives—the implosion method—was only beginning to show promise. Robert Oppenheimer had brought Kistiakowsky to Los Alamos to work on implosion. I was to be his right-hand man.

I was pleasantly surprised to find that I had been booked into a drawing room for the trip by train to Santa Fe. My best previous accommodation had been a compartment. I was further pleased to find that I shared my drawing room with the theoretical physicist Edward Teller. Edward was one of my scientific heroes, and although I didn't know him well, I had great respect for both his scientific abilities and his personal qualities. Most of the day and a half we traveled together, he spent tutoring me on nuclear explosives and shock-wave theory. I had not yet seen Bob Serber's *Los Alamos Primer,* notes on the lectures Bob gave when Los Alamos opened that summarized what was known up to that point about the nuclear and shock physics that would determine nuclear-weapons design. I had never even read anything about shock waves. Edward's one-man lecture course began my education.

Los Alamos—the Hill, we called it—was beautifully sited. From crossing New Mexico by car and train, I had imagined the state to be desert. To the contrary, the Los Alamos "mesas" were high, cool, scenic plateaus projecting eastward out of the mass of the Jemez Mountains, separated by narrow canyons. Vegetation was abundant, and much of Los Alamos was forested.

Initially I lived in the Big House, a rough-hewn dormitory a short distance from Fuller Lodge, the center of the Hill's social activities. After a few days I moved into the lodge in a room next to Kistiakowsky's. George was certainly the most competent high-explosives expert in the United States. He had been director of the NDRC explosives laboratory at Bruceton, Pennsylvania, and was unhappy at first to be sent into the boondocks to work on a program he thought could not help win the war. He soon changed his mind. He was a tall, vigorous Ukrainian, and the contributions he made were vital to the bomb program's success.

I set up office with him in the Tech Area, our desks pushed together so that we faced each other as we worked. He used me to serve as his eyes and ears to keep him abreast of the laboratory's research as it related to implosion. I attended information meetings as his representative and spent hours in the library reading reports on the work accomplished since the lab had opened, in April 1943.

During those first weeks George taught me about explosives. Since I had no experience in the field, it was a challenge for me to get up to speed.

High velocities imparted by guns or high explosives are necessary to assemble some, but not all, nuclear weapons. If no background neutrons intruded, one could simply take half a critical mass of nuclear material, push it slowly against something more than half a critical mass, and then inject a few neutrons. That would start fission going, and the material would explode. In the presence of a substantial neutron flux, however, the assembly must be completed before a single neutron gets into the nuclear material. Otherwise it begins fissioning prematurely and fizzles.

Emilio Segrè had moved to Los Alamos. He and his collaborators, including the young Owen Chamberlain, had measured the number of neutrons produced in the spontaneous fission of U-238 and found it to be small (U-235 doesn't emit any spontaneous neutrons, but the uranium Oak Ridge was preparing wasn't pure U-235). The muzzle velocity planned for the uranium gun bomb, Little Boy, made it unlikely that a neutron would enter the uranium pieces, bullet and target rings, before they were completely assembled. Each fissioning atom adds its 200 MeV to the whole; the more fissions, the more total energy released. At maximum criticality, neutron multiplication is fast and inertia holds the system together long enough to yield high energy. If a neutron enters the system before the assembly is complete, neutron multiplication starts more slowly; the system has time to heat up; it expands into an uncritical stage before much energy has been released; thus a fizzle, predetonation.

With modern weapons-grade uranium, the background neutron rate is so low that terrorists, if they had such material, would have a good chance of setting off a high-yield explosion simply by dropping one half of the material onto the other half. Most people seem unaware that if separated U-235 is at hand it's a trivial job to set off a nuclear explosion, whereas if only plutonium is available, making it explode is the most difficult technical job I know. I believe this distinction explains the preemptive Israeli attack on an Iraqi research reactor several years ago. That reactor used weapons-grade uranium, and I can understand why Israel was unwilling to allow such material to remain in the hands of so unfriendly a neighbor. That the reactor would make plutonium was, it seems to me, irrelevant. *All* reactors make plutonium; I doubt if Israel was worried about Iraq's developing a plutonium bomb. Given a supply of U-235, however, even a high school kid could make a bomb in short order.

When I arrived at Los Alamos, Joe Kennedy, a codiscoverer with Glenn Seaborg of plutonium, was directing a large chemistry laboratory devoted to the purification of plutonium. Since the half-life of plutonium is some twenty-five thousand years, it emits many more alpha particles per second than U-235, the half-life of which is nearly a billion years. Beryllium and other light-element impurities in plutonium react with its alpha particles to produce neutrons. Joe and his colleagues had learned to remove those impurities from plutonium down to a few parts per million, in order to reduce the neutron background and thus the possibility of predetonation. Even so, the planned plutonium gun would have to operate at what were then extremely high and problematic muzzle velocities to avoid a fizzle.

The plutonium Segrè studied first had been made in the Berkeley cyclotron. As plutonium samples began arriving from the reactors at Oak Ridge and, later, Hanford, Emilio found spontaneous-fission yields rising at an alarming rate. Enrico Fermi finally identified the problem: the longer the plutonium 239, the weapons material being bred from natural uranium, remained in the reactor and the larger the reactor's neutron flux, the more it would absorb neutrons and turn into Pu-240, an isotope with a very high rate of spontaneous fission.

Los Alamos was fortunate to have a full complement of high-caliber nuclear physicists. Because it did, a physicist of Emilio's skill could be diverted from apparently more essential tasks to measure the rates of spontaneous fission in samples of uranium and plutonium, constants we thought we already knew. His discovery of the high spontaneous-fission background of reactor-bred plutonium was the most important single event at Los Alamos in the first months after I arrived. It had fateful consequences for bomb design and for the work of Kistiakowsky's X division.

The conventional wisdom held that neutron capture could not compete with neutron-induced fission in the same isotope. Textbooks continue to record how Niels Bohr correctly concluded that fission induced in natural uranium by slow neutrons occurred in the rarer U-235 isotope, 1/140 the content of natural uranium, and capture and transmutation in the more plentiful U-238 isotope. It followed that U-235's cross section for fission had to be extremely high, given its extremely low natural abundance. Fission, Bohr argued, was a rapid process, occurring in about 10^{-20} seconds. Capture, on the other hand, took perhaps 10^{-15} seconds, the time required for the gamma ray to be emitted that stabilizes the nucleus afterward. When U-235 absorbed a neutron, Bohr thought, it would split before it had time to emit a gamma ray. So Bohr convinced everyone that no appreciable U-236 or Pu-240 could be made.

I was still in Chicago when the first evidence came along to contradict Bohr's universally accepted theory. Two graduate students using the Columbia cyclotron had built a much-improved version of my original neutron time-of-flight spectrometer. They measured the total neutron interaction cross sections for U-235 and U-238 by means of samples the Manhattan Project supplied. Apparently they didn't realize how startling their results were. I was with Fermi when he saw them. He was absolutely baffled, and he agonized over their explanation. A plot of the two isotopes' cross sections versus neutron energy showed the expected resonance absorption peaks in U-238—indicating neutrons captured to make U-239—but it also showed similar narrow peaks in the curve for U-235. Such narrow energies implied longer reaction times. Enrico was forced to conclude that capture and fission competed in U-235 (and in Pu-239, to which U-239 eventually decays).

One result of the discovery that Bohr had been wrong was a change in the calculated critical masses of uranium and plutonium needed to make nuclear weapons. The masses had to be increased to compensate for the decreased efficiency of fission, much to Ernest Lawrence's consternation—his calutrons, which would supply 99 percent of the uranium for the Little Boy bomb, would have to run longer now for a less effective return.

But the more grievous change the discovery forced on Robert Oppenheimer was the abandonment of the plutonium gun, which could not possibly operate at a muzzle velocity high enough to overcome predetonation in Pu-239 contaminated with Pu-240. And that was where George Kistiakowsky and implosion came in.

These important measurements of the physical properties of nuclear materials reached the Russians through spies with access to Manhattan Project secrets. Their reports effectively multiplied the Soviet Union's physics manpower. The USSR had very little nuclear-physics capability at the time, as I was aware from my extensive study of the physics literature. Without the Manhattan Project data, Soviet physicists probably wouldn't have discovered either the competing capture processes or the high spontaneous-fission rate of reactor plutonium until they had tried to build both uranium and plutonium bombs with incorrectly calculated critical masses and failed. They had no spare Segrès to waste on experiments the outcome of which they were sure they already knew.

The magnitude of this advantage came forcibly to my attention in England shortly after the war when I was having dinner with Sir William Penney, then head of the British Atomic Energy Establishment. Bill was properly upset with the U.S. Congress for reneging on the Roosevelt-Churchill agreement to share nuclear data after the war. In particular, he had been unable to find out what crucible material Los Alamos had used for the melting of plutonium, something he could have learned in five minutes in the Los Alamos library.

Without this small piece of information, which until the end of the war had been legally his, he had been forced to squander half a dozen man-years of his best metallurgists' time, so far without result. Plutonium, a highly reactive metal when melted, dissolved or interacted with all the crucible ceramics the British had tried. My position was difficult, because I knew the answer but couldn't tell Bill. During the war I frequently violated security regulations by passing along to American citizens information that I believed important to the war effort. I didn't feel I could violate the corresponding security regulations with Bill, a foreign national, even though the information was something he formerly had the right to know. I had to play dumb. But I saw at first hand the advantage the Russians won by spying.

Not long after I began working at Los Alamos, Robert Oppenheimer, obviously pleased to see me finally on his staff, assigned me to his steering committee, which governed the laboratory. It was composed of division heads and others like myself who had wide experience in war projects. Remembering the unworldly and long-haired prewar Robert, I was surprised to see the extent to which he had developed into an excellent laboratory director and a marvelous leader of men. His haircut almost as short as a military officer's, he ran an organization of thousands, including some of the best theoretical and experimental physicists and engineers in the world. The laboratory's fantastic morale could be traced directly to the personal quality of Robert's guidance.

General Groves had originally imagined that the scientists on the Hill should be inducted into the Army and awarded ranks appropriate to their ages and achievements. Robert had approved the arrangement, but men who came to Los Alamos from the MIT radar lab quickly changed his mind. It wouldn't have worked. Bob Wilson and Joe Kennedy, for example, who were under thirty but led the experimental nuclear-physics and chemistry divisions, would have been unable to argue with more senior but less knowledgeable officers. I don't think science can be done under authoritarian arrangements. And young people very often know better than their elders.

One program Robert inaugurated that boosted morale was a weekly evening colloquium that any properly cleared technical person could attend and where anything and everything about the secret work of Los Alamos could be discussed. What General Groves called compartmentalization—the restricting of information to those who needed to know—was the order of the day everywhere else in the

Manhattan Project. The benefit of getting the weapons built quickly, we felt at Los Alamos, far outweighed any possible damage from the escape of secret data. I'm sure the weekly colloquia provided Klaus Fuchs with much of the information he passed on to Moscow. I'm no less sure that they also saved countless American and Japanese lives.

Los Alamos was stimulating. I greatly enjoyed working there, meeting old friends, and making new ones. One of my early projects was finding a place for my family, which with Gerry's pregnancy now numbered three and a half. Housing was a continual problem on the Hill, where the number of residents doubled every nine months until the end of the war. Before the Army bought it, Los Alamos had been a private boys' school. Early arrivals with high status occupied the faculty houses, known collectively for their singular convenience as Bathtub Row. The Army built apartments in barracks style with tin showers. The apartments came in various sizes, the commonest being four-family units, with two families above and two below. All four units shared the same furnace, which had no thermostat and burned smoky coal. In the winter the Chicano furnace crews stoked them full blast; to control the temperature in our rooms, we opened and closed the windows.

Fortunately for us, an old friend, Vera Williams, ran the housing office. She assured me I could have the next available apartment. When the three other wives in the building where that apartment came available learned that they were to have a neighbor named Alvarez, they rushed to Vera to complain of their imminent loss of social standing—they certainly didn't count Spanish-Americans among their best friends. Vera calmed their fears. A month after I arrived, I returned to Chicago to pick up my family. Gerry was happy to renew many friendships. Quite a few Los Alamos wives had children about Walt's age, and quite a few were pregnant. Walt went to nursery school. In the evening he and I often walked together examining the new buildings that were going up. One of my Army construction friends nicknamed him the Chief Building Inspector.

The Tech Area was a long walk from our apartment. For the first several months I drove my jeep back and forth. Then a new regulation made the use of government transportation for private purposes a federal offense. I bought a bike. We saved our gasoline ration for occasional trips to Santa Fe, visits to Indian ruins, or tours up to the Valle Grande, the spectacular volcanic caldera above Los Alamos to the west, of which the mesa itself was eroded tufaceous spill.

In the winter I could ski on a run a few thousand feet above the Hill on Sunday morning (we worked six days a week) and then come home, collect my family, and picnic in warmth down in one of the canyons. Kistiakowsky and Hugh Bradner, two of the best skiers we had, laid out a ski run and installed a rope tow that Army enlisted men operated in their spare time. George and Hugh knocked down the trees with Composition C—the putty-like explosive plastique—which was easier than sawing them down.

Army personnel ran all the town services—laundry, grocery store, hospital, post office, movie theaters, and so on. (Once, when I dropped into the University of Chicago barbershop on a visit, my old friend the barber said, "What's up, Doc—you cutting your own hair these days?") Unmarried staff ate in the cafeteria, families usually at home. Every few weeks a truckload of steaks would arrive, and the families would rush to the cafeteria. I played poker once a week with friends, including, on occasion, John von Neumann. Johnny, one of the world's great mathematicians, wrote the classic *Theory of Games and Economic Behavior* but usually left our poker games poorer than when he arrived.

My daughter Jean was born four years to the day after Walt. Two years previously Gerry and I had lost the first Jean Alvarez at birth under traumatic circumstances. Gerry had mentioned that her obstetrician showed no interest in her as a person. How cold the man was I learned when he appeared in the waiting room and announced abruptly that the baby had died. He volunteered nothing about the circumstances and offered no condolences. Having delivered his message, he turned on his heel and walked away. It was a sad and difficult time for both of us. By contrast Jim Nolan, the Army obstetrician at Los Alamos, was a wonderful person. I got to know him better when I was ill for a time with hepatitis. How much I valued him is perhaps measured by my decision, in the Pacific just before I flew with the first atomic bomb over Japan, to entrust him to deliver a letter to Gerry and the children if I didn't return. I knew he would deliver it in a way that would make me happy.

Gunpowder, a Chinese invention, is an example of a low explosive. High explosives, which have great shattering power—brisance—are characterized by the high velocity of the chemical reaction's detonation front. An artillery shell is driven out of a gun barrel by a low explosive (a high explosive would shatter the barrel) and contains in its body a high explosive, which detonates after impact. The basic high-explosive reaction is that of a solid turning into a high-pressure, high-temperature gas, the shock wave from which drives everything before it.

Alfred Nobel devised the first effective method for detonating high explosives. Until I worked on the problem, a detonator always fired a small charge of a sensitive explosive like lead azide or mercury fulminate. In those materials a small pressure wave develops into a high-speed detonation wave that can then be sent into an insensitive high explosive like TNT. Practical high explosives are so insensitive to shock that they must be accelerated to thousands of times the force of gravity (which is what the detonation wave does to them) to explode. Crystals of TNT fall everywhere in TNT factories: on the floor, where they are walked on, and on rails, where shunting railroad cars pulverize them without effect. World War I was fought largely with TNT, often diluted with ammonium nitrate. Chemists developed two important new high explosives between the wars, RDX and PETN. RDX, the main explosive of World War II, was diluted with TNT in different proportions to form Composition B and C. We used Comp B at Los Alamos in the plutonium implosion bomb, Fat Man. PETN is a powdered high explosive; it fills the core of the commercial fuse known as Primacord. The detonation velocity in RDX and Primacord is some five miles a second—eight millimeters a microsecond.

The easiest way to start a nuclear explosion, as I've said, is to bring two subcritical pieces of fissionable material quickly together in a gun to make a critical mass. A critical mass is not some absolute number of kilograms of fissionable material. The amount of fissionable material to be used depends on the density of the material chosen and its shape. A mass that is critical at the density of cold metal becomes subcritical when its fissioning atoms release enough energy to heat it and expand it to lower density. At some point the chain reaction stops.

The method that George Kistiakowsky was working to perfect, implosion, reversed this process: it involved compressing a subcritical mass of fissionable material to higher density with shock waves from a surrounding high-explosive shell, squeezing it to supercriticality. The immediate and obvious advantage of implosion is that the material "assembles" so quickly under the pressure of the high-speed shock wave that it doesn't have time to predetonate.

Seth Neddermeyer invented the implosion method at Los Alamos shortly after the laboratory opened in March 1943. Neddermeyer was Carl Anderson's coworker at Caltech in 1932 when Anderson discovered the positron. A few years later he was senior author with Anderson of the paper announcing the discovery of the muon.

Seth's original idea was to shape the fissionable material as a hollow sphere, which a high explosive wrapped around it would implode into a solid supercritical mass. That arrangement was too difficult to configure during the war. It has some of the properties of a shaped charge, a block of high explosive with a conical depression lined with a thin coating of metal. When the detonation wave of a shaped charge hits the liner, it compresses it into a high-velocity jet of liquid metal that can penetrate up to a foot of armor plate. When I arrived at Los Alamos, I found people trying hard to overcome the jetting problem, which caused test cores in the bomb to squirt out uselessly and effectively disassemble. They had very little success. Jets formed when the detonation waves from the various detonation points spaced around the high-explosive shell came together. We needed computers to manipulate the complicated spherical geometries. They hadn't been invented yet. I'm confident the weapons laboratories of the world, with their major computer installations, solved the problem long ago.

The Fat Man bombs—the one tested at Trinity Site, north of Alamogordo, and the one exploded over Nagasaki—used a somewhat different implosive configuration proposed by Robert Christie, two solid hemispheres of plutonium metal, just subcritical, compressed to greater density by a surrounding five-foot shell of high explosives. At first the lab tried to develop a solid high-explosive shell. Mistimed detonation on the surface of the shell still caused jetting, detonation waves arriving at different intervals and interacting as before. Moreover, the waves did not have the right geometry to squeeze the sphere of plutonium uniformly smaller.

To make implosion work, we thus required two inventions. The first was the explosive lens, which behaves much like an ordinary glass lens: it takes a spherical wave that is expanding from a point and turns it into a spherical wave that is converging on a point. The second invention was simultaneous detonation. I led the team that made the second invention.

I learned a lot about the five regular polyhedra during this period. The largest number of points that can be spaced equally on the surface of a sphere is twenty, a number corresponding to the centers of the twenty triangular faces of an icosahedron. The next-largest number is twelve, corresponding to the centers of the twelve pentagonal faces of a dodecahedron. It's possible to interleave a dodecahedron with an icosahedron, as Plato showed, to get the nearly regular faces, alternately pentagons and hexagons, that shape an object so familiar as a soccer ball. The thirty-two points became, in spherical implosion, the thirty-two points of detonation. Both before and after the invention of explosive lenses, much effort went into firing those thirty-two detonations simultaneously.

The method under development when I arrived at Los Alamos used a single electric detonator that started a detonation wave moving down a length of Primacord. The Primacord then branched like a tree into thirty-two segments of exactly the same length. That required a total of about a hundred meters of Primacord inside the bomb case, an awkwardly bulky harness. Primacord was necessary because electric detonators, which Nobel invented and developed, had timing uncertainties of about a millisecond. Initiate an explosive charge with thirty-two of the best detonators then available, which had a half-millisecond spread, and the detonation waves would arrive at the nuclear core spaced by distances of ten or twenty feet rather than the required fraction of a millimeter. The millisecond spread resulted from slight irregularities in the sensitive explosives of the detonators themselves. It caused the timing of the conversion of the initial electric pulse into a detonation wave to vary. No one therefore contemplated using more than one detonator.

But the bulky Primacord had problems of its own. The density of its powdered PETN varied from point to point along its length, and this varied the detonation velocity. No one was confident that the manufacturer could achieve the quality control necessary to overcome this difficulty. "Why use lead when gold will do?" was the motto at Los Alamos. We threw a lot of money at the problem without success.

By this time I had learned just enough about high explosives to experiment with them but not enough to be sure that an idea wouldn't work. One day I suggested to George Kistiakowsky the possibility of starting the detonation wave in a detonator by discharging a high-voltage capacitor (a device for storing electricity) through the detonator's bridge wire. George was skeptical but agreed I should give it a try.

Larry Johnston, my former student and coinventor of GCA, had just arrived at Los Alamos from MIT. He prepared to try out my idea. He'd never seen an explosive fired before, so I introduced him to the technology. His first experiment used a one-foot length of Primacord with a detonator taped to each end and a block of lead taped to the middle. Larry discharged a fifteen-kilovolt, one-microfarad capacitor through the two detonator bridge wires in series, thus starting detonation waves moving in the opposite ends of the Primacord toward the middle. When the two waves collided, the increased pressure would dig a crease into the lead. If the detonators fired with perfect simultaneity, the crease would show up at the center of the block face. If one detonator started ahead of the other by one microsecond, the crease would be off center by four millimeters, half the distance either wave would travel in a microsecond. (If two detonators had been set off in the usual way, by burning out their wires with a low-voltage source, the difference might have been as much as ten feet.)

The detonators in Larry's first test fired simultaneously within less than one microsecond. In a few days he was firing independent electrical detonators with a timing spread of less than one-tenth of a microsecond, at which point he changed to an aluminum block for more precise time measurements. In a few months my group, now abruptly enlarged, brought the timing spread for hundreds of detonators fired simultaneously to within a few billionths of a second.

We were much too busy to pursue ordinary commercial applications at the time, but my hunch that explosives could be fired by discharging a high-voltage capacitor through wire has revolutionized detonation technology. What Larry and I had stumbled upon is common knowledge today: that such a discharge causes a wire suddenly to vaporize—to explode—and that the rapid expansion of the vaporized material is indistinguishable from a detonation wave. We demonstrated that we could detonate such insensitive explosives as TNT, PETN, and RDX directly with exploding wires and with no intervening sensitive explosives like mercury fulminate as couplers.

Now we had to develop our new detonators and devise methods for studying the time spread in hundredths of a microsecond between groups of thirty-two and even sixty-four detonators fired simultaneously. The Army quickly erected a large building for us on South Mesa, across Los Alamos Canyon from the main Tech Area, and we assembled a sizable group of technical people from the Special Engineering Detachment (SED), men with technical and sometimes scientific training who were drafted and sent to Los Alamos to help with the weapons program.

We solved the timing study problem with the help of Ira ("Ike") Bowen, Robert Millikan's longtime colleague in cosmic-ray research and, later, the director of the Palomar Mountain Observatory. Ike was a skilled optical designer who developed a streak camera for the Navy that relied on a high-speed rotating mirror of the kind Albert Michelson had used to measure the speed of light (a streak camera records photo finishes at horse races). To check the firing simultaneity of thirty-two detonators, we mounted them one above the other in a column. If all the detonators fired simultaneously, the resulting streak film would show thirty-two bright spots in a line. If one of the detonators misfired, its spot would lag behind.

Charlie Lauritsen of Caltech, who had been developing accurate rockets for the Navy, urged me to visit him in Pasadena. I toured his impressive rocket laboratory there, and he put me in touch with manufacturers who could produce our new detonator bodies—two heavy wires imbedded in plastic with hairlike bridge wires soldered to connect them. I ordered thousands of units. I made one serious error: I economized on the wire that linked the bridge wire to the firing circuits. I wasn't entirely at ease yet with the fact that war is unbelievably wasteful. We had trouble with those wires; the insulation tended to short out.

Tom Lauritsen, Charlie's son, persuaded me on a second trip to Pasadena to forget about cost and buy the most expensive materials available because the speeding of development in wartime and the improving of reliability saved lives. My group then designed and ordered tens of thousands of detonators each strung with some ten feet of the most expensive coaxial cable on the market, which would be blown up when the detonators were set off.

Our South Mesa operation was different from any I had ever been associated with before. Some of our Army men were technically skilled; others served us as laborers. A large force of Spanish-American women imported from nearby towns loaded detonators behind armored shields. We used insensitive explosives and lost not so much as a finger. We experimented to decrease the timing spread, sorted different particle sizes, varied packing pressure, changed all the engineering variables we could think of. It was cookbook work more than scientific, but we knew it was important and pushed it hard.

While we tested detonators, Kenneth Bainbridge's group adapted our system to Fat Man. His deputy, Don Hornig, later to be President Lyndon Johnson's science adviser, engineered the X unit, the high-voltage capacitor-based system that simultaneously fired all the Fat Man detonators.

When the detonator development was sufficiently advanced, I moved into the physics division building to work under Bob Bacher. My office mate there was Charlie Critchfield, a theoretical physicist from Harvard who was designing the beryllium-polonium initiator that would release a burst of neutrons at exactly the right moment to start fission in the Fat Man implosion bomb. Since the spontaneous fission of Pu-240 provided a flood of neutrons, this device may not have been necessary, but we developed a number of such contingency devices to make sure our bombs worked. Niels Bohr, who visited Los Alamos for months at a time, found the problem of the initiator fascinating and dropped by daily to chat, a fringe benefit for me of sharing the office.

In the fall of 1944 Robert Oppenheimer asked me to supervise what came to be known as the RaLa experiment. Radiolanthanum is a decay product of barium 140, which in turn is a product of nuclear fission. Ba-140 decays to radiolanthanum with a thirteen-day half-life; radiolanthanum further decays with a forty-hour half-life and the emission of copious quantities of high-energy gamma rays. The intermediate-scale reactor at Oak Ridge made large amounts of fission products; I arranged for the chemists there to separate out a sample of radiobarium equivalent to several hundred curies of radium, an intensely radioactive source (probably the strongest source ever prepared up to that time). We had the radiobarium trucked to Los Alamos in a heavy, lead-lined cask. In a specially built separation facility we milked the radiolanthanum from its barium parent. We wanted to measure how effectively implosion squeezed a shell of material to greater density. The RaLa experiment proposed to do so by loading radiolanthanum at the center of a hollow metal sphere and surrounding the test unit with gamma-ray detectors. As the explosives squeezed the sphere, its thickness would increase. That would block more gamma rays, and the gamma-ray count would decrease. A few microseconds later the whole system would be blown to smithereens.

I didn't construct or operate the equipment, but I was effectively project coordinator and experiment designer, assigning jobs and making sure that everything was ready when it was supposed to be and that the experiment worked. My assistant Hugh Bradner and I set up shop in the canyon just north of Los Alamos. Bruno Rossi, my candidate to make the gamma-ray detectors, did a remarkably good job. The explosive charge and sphere were built nearly full-scale. I designed the chemical separation equipment, conveniently installed in a truck, and the source manipulating gear that allowed us to move the RaLa to the sphere without excessive exposure to its gamma radiation.

We needed to be close enough to the gamma-ray detectors to record their signals but shielded from the blast. I visited the Dugway Proving Grounds in Utah and ordered two Army tanks with their guns removed. I asked Kistiakowsky to calculate how close we could park the tanks to the explosion—hundreds of pounds of RDX, hundreds of curies of gamma rays—and still be assured of survival. To make certain George checked his calculations, I invited him to join us in one of the tanks, which he did. I was looking through the periscope when we fired the shot, but the shock wave blew dust into my eyes. When the air had cleared, we climbed out of our tanks to find the woods on fire. White-hot metal fragments had ignited the trees and the underbrush. We weren't in danger from radioactivity—it was too widely dispersed—but the fire that almost completely surrounded us was a source of worry. We managed to contain it. We could always have retreated back into our tanks.

The gamma-ray records proved less definitive than we had hoped but helped develop the implosion method. I cleaned up some loose ends in detonator design. By April 1945 that work had moved on to engineering, and I was again unemployed.

I went to see Robert Oppenheimer and asked him if he had any assignment that might take me to the Pacific. As a matter of fact, Robert said, he had just the job for me, but I wouldn't have much time to make it work. He'd just realized we had developed no method to measure the energy released by the bombs we were preparing to drop on Japan. Normally a new weapon is proof fired at a proving ground before it's used in combat. We had only one U-235 weapon, however, every atom of which had been run twice at enormous expense through the Oak Ridge calutrons, and as of July 1945, there would be about one plutonium bomb a month, the first of which would have to be used up in a static test to make sure the Fat Man implosion system worked as we predicted it would. Since we couldn't proof fire our bombs, then, Robert proposed that I work out a way to transport a proving ground over enemy territory to measure the energy release of the bombs in combat. Ken Bainbridge's people were then instrumenting the test shot in the desert at Trinity Site, two hundred miles south of Los Alamos. Robert suggested I talk to them to determine what measurements might be adapted to combat conditions.

The most definitive tests planned for Trinity—the measuring of neutrons, gamma rays, and the rate of expansion of the nuclear fireball—couldn't be repeated in combat. The bombs were to be dropped from B–29's flying at 32,000 feet to detonate at about 1,900 feet; neutrons and gamma rays would be completely absorbed by the intervening air. I couldn't think of a reliable way to measure the fireball expansion. At Robert's urging, filming was tried at the last minute. Attempts were made with a high-speed camera from twenty-five miles away; they failed both times, which perhaps justified my pessimism.

I decided that the obvious approach was to measure the atmospheric blast pressure. I would need a microphone calibrated so that its signal would increase as the blast wave hit it and decrease in response to the negative pressure as the rarefaction wave passed by. Bill Penney calculated the expected pressure at a microphone suspended by parachute at 30,000 feet near the bomb drop point. I looked through acoustics texts without finding a suitable way to measure such pressures. Then I remembered that in Pasadena a few months earlier I had read a report about a device that might meet my needs.

Caltech's Jesse DuMond, a talented physicist with exceptional ability at instrument design, had written the report. It concerned a mechanism for improving the accuracy of antiaircraft gunners, who in training often missed their towed canvas targets entirely and had no way of knowing how to adjust their aim. Jesse had suggested mounting a module containing two back-to-back microphones in the target sleeves. The shock waves from the supersonic bullets would register on the two microphones; the distance of the miss would be related to the sum of the signals; the azimuth angle of the miss would be related to the difference between the signals. This information could be fed back to the gunner in real time, and he could then correct his aim. The firing error indicator that Jesse designed did the job, but by the time it was ready the development of aiming computers, radar, and proximity fuses had mooted the problem.

Reading Jesse's report, I had seen for the first time the N wave that characterizes a bullet's passage, the air pressure rising suddenly to a high positive value, decreasing linearly to a nearly equal negative value and then abruptly returning to zero. Jesse's invention used a frequency-modulated transmitter to relay this information to a ground receiver. A shock wave pushing on the microphone diaphragm changed the signal frequency, and the ground receiver converted this change into a voltage-versus-time signal that reproduced the N wave. I confirmed that the shock pressures from bullets passing within some tens of feet were comparable to the shock pressures anticipated from an atomic bomb at a range of eight miles. I knew then that Jesse had something that could be of great value to me and arranged to visit him to discuss obtaining some of his excellent microphones and their associated FM transmitters and receivers.

Most of the work on the system, Jesse told me, had been done by his new son-in-law, a young German émigré physicist named Pief Panofsky. I asked security to clear them both so that I could tell them my plans for their microphones.

When I got to Caltech, I found that Panofsky was off at a remote Army base in the Sierra Nevada demonstrating his system. Jesse, who had spent the war on the sidelines and was delighted that one of his creations might now play a useful role, drove me into the mountains to meet his son-in-law. I knew that Jesse was wealthy and cosmopolitan and expected the man who married his daughter to look like a German nobleman. Pief's appearance fitted my preconception not at all. He was short and dark and rotund and so extraordinary a physicist that he came to be my secret weapon at Los Alamos and at Berkeley after the war; he recently retired as director of the Stanford Linear Accelerator Center, the great laboratory he founded. By any definition he is now known as one of the greatest physicists of the postwar era.

I described a plan to calibrate the parachuted microphones after they reached their free-fall velocity. The oscilloscope record radioed from the microphones to the B–29 would show first a calibration pulse of known pressure and then the blast-wave pulse from the exploding bomb. Pief thought my idea workable, said he would start working on it as soon as he returned to Pasadena, and gave me his spares.

At Los Alamos I quickly put together a group of young physicists whose earlier work, like mine, was finished: Larry Johnston; Bernie Waldman from Notre Dame; Harold Agnew, who would go on to become the third director of Los Alamos, after Robert Oppenheimer and Norris Bradbury. I ordered up half a dozen SEDs (technically skilled GIs) to operate the instruments over Japan—Robert had said we wouldn't be allowed to fly in combat ourselves. I quickly realized I wanted to be on hand when these measurements were made and started a campaign, eventually successful, to allow Larry, Harold, and me to fly with our equipment.

We altered the transmitters to give them more range and adapted them to battery operation. Pief's microphones had pointed outward from the surface of an eight-inch hemisphere. We reoriented one of them to point downward. Above it was the power amplifier and the battery pack. The whole assembly fit into an aluminum cylinder eight inches in diameter and three feet long. Four eye bolts attached a parachute harness. Our detectors would be dropped from the bomb bay of a following B–29 at the same time the bomb was dropped. The two bombers would then get the hell out, the instrument plane in due time receiving the radio signal.

By the end of April I was ready to find a manufacturer. I called on my friends at the Gilfillan Company in Los Angeles who were making GCA radar units. They set up a production line and turned out a few dozen bomb detectors in less than a month.

The B–29's had to be fitted with bomb bay racks for the detectors and the receivers installed with their associated oscilloscopes and recording cameras. Harold, Larry, and I flew up to Wendover Field, on the Nevada-Utah border, to visit the 509th Composite Group, Paul Tibbets's outfit training there to drop the first atomic bombs. It was a great pleasure to be back with pilots and aircraft after a long hiatus. We flew with a bombing crew to the Salton Sea to observe a practice drop of a dummy Fat Man, our first B–29 ride, and returned to Los Alamos with detailed drawings of the big bomber's rear compartment.

Pief arrived with the modified microphones. He was officially a consultant, but as soon as he walked into our laboratory he peeled off his coat, grabbed a soldering iron, and began making wiring changes in the receiver to adapt it to our needs. Enthusiastic participation has become Pief's trademark; it made me decide then and there to hire him once the war was over. He could become my secret weapon because no one else in the old guard knew he existed. After the war all the bright young men who had surfaced within sight of the senior men at MIT and Los Alamos ended up with good university jobs; the unknown Pief accepted a job at Bell Labs, but before we lost him to New Jersey I brought him to Berkeley.

When the receiver was adjusted, Pief taught us how to service the equipment down to the smallest detail. We needed to know; we would be operating it far from technical support. Testing and installation followed, and then I planned a test flight. Most of my colleagues hugged the ground at dawn on the morning of July 16, 1945, when the first atomic bomb ever exploded on earth was detonated on a hundred-foot tower at Trinity Site. I was in the air.

Our B–29 was supposed to fly directly over the tower and drop our parachute gauges just prior to the firing. For safety we equipped the plane with radar beacons and made sure an SCR–584 radar tracked us from the ground. The tracking data were relayed to the Trinity control room so that the firing crew would know at all times where we were.

I left Los Alamos on the afternoon of July 15, telling Geraldine what most Los Alamos wives were told, that if she happened to be up at five o'clock the next morning and looked south she might see something interesting. I drove down to Albuquerque and had dinner with Ernest Lawrence, who was there, like me, to watch the shot. After dinner I drove to Kirtland Air Force Base, where the B–29 was waiting. Late that evening I learned that Robert Oppenheimer was trying to reach me.

I called him from a public telephone at the Albuquerque airport. He told me he had decided that we should fly no closer to the tower than twenty-five miles. We were scheduled to leave for the Marianas in less than a week; Robert's decision meant we couldn't check our detector system before we left. I explained in great detail our arrangements interlocking the ground radar tracking system with the Trinity control room that assured our safety. Robert was insistent and ordered me to abandon the test. I was absolutely furious, angry with him as I have never been angry with anyone before or since, but I had to back down because he was my boss. (My anger quickly subsided. Robert was under tremendous pressure that night for reasons none of us knew: the President of the United States was waiting impatiently at Potsdam for word of the test.)

Sometime after midnight we took off. Larry, Harold, and Bernie were aboard, and Pief was our guest. The plane also carried Navy Captain William ("Deke") Parsons, who headed the Los Alamos ordnance division and would fly with the Little Boy bomb. We tuned in the Trinity control room on our radios, circled in the stormy night twenty-five miles from the tower, and followed the progress of the countdown. As it came to its final moments, the pilot banked and headed toward the tower.

Shortly after the Trinity test I prepared an eyewitness account of what I had seen. This is what I wrote at that time:

> I was kneeling between the pilot and copilot in B–29 No. 384 and observed the explosion through the pilot's window on the left side of the plane. We were about 20 to 25 miles from the site and the cloud cover between us and the ground was approximately 7⁄10. About 30 seconds before the object was detonated the clouds obscured our vision of the point so that we did not see the initial stages of the ball of fire. I was looking through crossed polaroid glasses directly at the site. My first sensation was one of intense light covering my whole field of vision. This seemed to last for about ½ second after which I noted an intense orange-red glow through the clouds. Several seconds later it [seemed] that a second spherical red ball appeared but it is probable that this . . . phenomenon was caused by the motion of the airplane bringing us to a position where we could see through the cloud directly at the ball of fire which had been developing for the past few seconds. . . . In about 8 minutes the top of the cloud was at approximately 40,000 feet as close as I could estimate from our altitude of 24,000 feet and this seemed to be the maximum altitude attained by the cloud. I did not feel the shock wave hit the plane but the pilot felt the reaction on the rudder through the rudder pedals. Some of the other passengers in the plane noted a rather small shock at the time but it was not apparent to me.

As I think this account makes clear, we experienced the shot not nearly so dramatically as people on the ground reported; at 24,000 feet we may have felt more detached.

Arthur Compton told of a lady who visited him after the war to thank him for restoring her family's confidence in her sanity. She had visited her daughter in Los Angeles and was driving home across New Mexico early one morning to avoid the midday heat. She told her family that she saw the sun come up in the east, set, and then reappear at the normal time for sunrise. Everyone was sure that Grandma had lost her marbles, until the story of the Trinity shot was reported in the newspapers on August 7, 1945.

After the test we drove back to Los Alamos, packed our footlockers, and flew to San Francisco, from which we staged for the B–29 base on Tinian, fifteen hundred miles from Japan.

EIGHT

After Hiroshima

AFTER FLYING the Hiroshima mission we found a party in progress when we landed on Tinian, but hardly anyone knew about the atomic bomb—the secret remained secret, even on Tinian, until President Truman issued the first public announcement. Instead of getting the traditional steak that bombing crews received when they completed a mission, we stood in line at the group picnic for hot dogs. A few of our scientific friends questioned us; we were never debriefed.

When we developed our oscilloscope film, we found we had obtained only one good record, but it was exceedingly good: we got the same value for the peak shock pressure using as a baseline either our in-flight calibration pulse or an absolute calibration we had measured the afternoon prior to the mission. Before we could calculate the bomb yield, however, President Truman announced from Washington, without evidence so far as I know, that the yield was equivalent to 20,000 tons of TNT. (In 1953 I learned from a theorist at Los Alamos, Fred Reines, that the damage done at Hiroshima couldn't be reconciled with the canonical 20 kilotons. I gave Fred the records from my personal files. With them he calculated an energy release of 12.5 KT, plus or minus 1 KT. Thirteen KT has now replaced the old value of 20 KT.)

When the President announced the bombing, Tinian was delirious. The 509th had been a black-sheep bunch to most of the other aircrews, commanding the best aircraft and the highest priorities and seemingly doing nothing more to earn them than drop odd-looking 10,000-pound bright-orange bombs (high-explosive dummies, nicknamed "Pumpkins," for practice). Now there was nothing but praise. Men who had been dreading the thought of the inevitable invasion took heart that the tools with which to end the war were at hand.

The Japanese didn't agree, it seems. The bombing was hardly discussed among the military leadership and was minimized in public reports. If we had not had another bomb available to drop very quickly, I'm quite sure, the Japanese would have kept on fighting, expecting to inflict terrible damage on us in the fall when we invaded their cherished homeland.

If I had been a Japanese nuclear physicist advising the war ministry, I would have told them how expensive and difficult it was to separate uranium isotopes—plutonium was still an Allied secret—and would have estimated that another such bomb as Little Boy probably wouldn't be ready for months. That would have been correct. In mathematical terms, the Nagasaki bomb was both necessary and sufficient to force the Japanese to accept unconditional surrender, which they were maneuvering even after Nagasaki to avoid.

Many people wanted to fly that mission as observers, including Bill Penney and Group Captain Leonard Cheshire, an RAF hero with a Victoria Cross. General Groves had refused to allow any British aboard at Hiroshima but relented for the second mission and put Penney and Cheshire in the Fastax plane. The other new crew member was to be Bob Serber; Bernie Waldman gave him a crash course in Fastax operation. Harold Agnew and I didn't fly the Nagasaki mission, but Larry Johnston did, making him the only person to have seen all three of the first atomic bombs explode. Quiet and deeply religious, Larry believed with the rest of us that what we were doing would save innumerable American and Japanese lives.

On the evening of August 8 Bob Serber, the theoretical physicist Phil Morrison, and I sat in the officers' club discussing the forthcoming mission and wondering what we could do to help shorten the war. I remembered that our former Berkeley colleague Ryokichi Sagane was a professor of physics at the University of Tokyo and suggested we send him a message. It would appeal to him to inform the Japanese military that, since two atomic bombs had been dropped, it was obvious that we could build as many more as we might need to end the war by force. I drafted a letter that Bob and Phil edited. Then I wrote out an original and two carbon copies, sealed them into envelopes, and taped them to the three parachute gauges to be dropped over Nagasaki the following morning. The letter was quickly found, delivered to the military, and discussed at length, but Sagane wasn't contacted and saw it only after the war. Through his courtesy I have the one surviving letter and the Panofsky pressure gauge to which it had been attached.

Nagasaki was not a primary target; the primary target announced at the briefing I attended on the night of August 8 was the large Japanese arsenal on the northwest coast of Kyushu at Kokura. Under the blackout conditions after the briefing, I went with Bob Serber to the equipment room to pick up his flight gear, delivered him to his plane, shook our several friends' hands, and wished them a good trip over the empire. Harold Agnew and I watched the planes take off from a hill south of the field, where I managed to give myself a thorough fright imagining I heard Japanese soldiers—some were still at large on the island—stalking us through the sugar cane. Back at the 509th area I stopped at my tent to find a dejected Bob Serber sitting on the stoop. There had been a parachute check at the end of the runway, Bob said, before his plane took off. He discovered he had picked up an extra life raft instead of a parachute, and the pilot had ordered him out—although Bob and his Fastax camera had been the only reason this third plane was flying.

That snafu proved to be prophetic of the Nagasaki mission, as abominably run a raid as any in the history of strategic warfare. Everyone connected with it must have been horrified by its confusion. The Fastax plane was late for rendezvous over Yaku-shima. The bomb plane, *Bock's Car,* piloted by Chuck Sweeney, had a transfer pump malfunction that reduced its fuel reserve by six hundred gallons. *Bock's Car* and the *Great Artiste* dallied too long over Yakushima waiting for the Fastax plane to find them. The weather in the meantime closed up over Kokura Arsenal. Once there, *Bock's Car* made a first, a second, and then a third run while the bombardier, Kermit Behan, searched in his bombsight for the target. Flak from the Yawata Steel Works to the west, one of the most heavily defended centers in Japan, began to find the bombers' altitude. Sweeney's fuel was critical by then, and he decided to fly on to Nagasaki. He was supposed to fly around Kyushu and approach the port city by sea but flew directly across the island instead. He would have to drop the bomb on the first pass in order to reach Okinawa, the closest base to Japan. The only—and unthinkable—alternative was to ditch it in the East China Sea.

Sweeney decided against orders to bomb if necessary by radar. Ostensibly a hole opened in the 80 percent cloud covered target in the last moments of the run. I've always taken this hole in the clouds with a grain of salt, since Behan, one of the best bombardiers in the Air Force, missed his target by two miles, a reasonable radar error in those days. Fat Man exploded above a narrower stretch of the Urakami River valley than it had been targeted for; its yield was twenty-two kilotons, but it caused less damage than Little Boy, though damage enough to serve its grim purpose. The Mitsubishi factory almost directly below the point of explosion had made the torpedoes that devastated Pearl Harbor.

Waiting on Tinian for the war to end, most of our Los Alamos contingent settled into recuperating from five years of hard work, sleeping late, swimming, enjoying our beer rations, watching movies, playing poker until well after midnight. The beaches were beautiful and the water warm. One of our pastimes was collecting cowrie shells, the traditional Pacific island currency. We dove for the shells wearing face masks, but the ones we found weren't shiny and still had cowries inside. It occurred to me that the war prisoners on the island might have supplies of cowries. When we shipped our electronic equipment to Tinian, we had included special expediting packages in the shipment. Mine was a case of whiskey, which can work wonders among GIs. Harold Agnew's was a large case of soap. An overpowering stench blew continually from the prison compound. I visited a quarry where the prisoners worked, and with vigorously applied sign language negotiated an exchange with their leader, soap for cowries.

I returned the next day with a dozen bars of soap. The prisoner presented me with bag of a hundred beautifully polished cowrie shells. We were both so happy with the exchange that we didn't haggle at all. In the hands of my colleagues the Cowrie Cartel grew apace. The prisoners earned more soap than they had seen in years, and we acquired an enormous number of cowries. I became known around the 509th as Trader Alvarez, the man who cornered the Tinian cowrie market.

After the surrender ceremony in Tokyo Bay, General Groves, who had kept us on station to make sure the Japanese conceived no last-minute change of mind, released us to go home. With our Manhattan Project clearances we were able to fly on the 509th's Green Hornets rather than wait for the slow, crowded troop transports. My flight to Hawaii was just as long and dull as the flight out had been, but an old friend from Berkeley and MIT, Lawrence Marshall, met my plane at Hickam with a staff car and a driver and announced that we were going to celebrate. Larry was a special aide on the staff of General Richardson, the commanding general of the Army in the central Pacific. He assured me I wouldn't miss my plane; the driver, he said, would maintain radio contact with headquarters.

We drove to a beach cottage north of Waikiki from which spilled a wild celebration. I was welcomed as a conquering hero: most of the people at the party had expected to take part in the invasion of Japan. After morning cocktails and a swim, we moved on to the Outrigger Canoe Club. I learned to surf. Larry organized a dinner in the club ballroom and arranged for me to escort a pretty Army nurse, the first woman I'd talked to in two months. Dinner led to dancing, which led to singing. Larry reported periodically that my flight wasn't nearly ready to go. We ate again and drank some more. We swam at midnight in the warm tropical sea. Then it was time to go.

My coach turned into a pumpkin. After a day of immense popularity I now became a pariah. When I arrived with a crowd of happy people decked in leis and was plied with last-minute champagne, my flight mates thought their plane had been delayed so long because of me (but in fact I wasn't the reason for the delay). I didn't have to confront their hostility for long. I slept the entire twelve hours to San Francisco.

My last weeks at Los Alamos were not happy. Many of my friends felt responsible for killing Japanese civilians, and it upset them terribly. I could muster very little sympathy for their point of view; few of them had any direct experience with war and the people who had to fight it. I've thought about the issue of using the atomic bomb for more than forty years now and kept my thoughts to myself. As part of this accounting of my life, it's time to express them.

The first question that concerns me is whether or not we ought to have demonstrated the bomb to the Japanese without loss of life. Ernest Lawrence, Arthur Compton, Enrico Fermi, and Robert Oppenheimer were asked to consider this question in June 1945. "We can propose no technical demonstration likely to bring an end to the war," they responded in a formal memorandum; "we see no acceptable alternative to direct military use." (All four of these men were close friends of mine, and each had been, at different times, my immediate superior. They all had exceedingly high moral standards. I believe their formal conclusion must be taken seriously.)

A demonstration, it seems to me, might have convinced a jury of Japanese scientists that the war was lost. But the war was run by professional soldiers, not scientists, and there is plentiful historical evidence that the Japanese high command was unimpressed. It's even probable that Japanese military leaders would have been uplifted by a decision to stage a demonstration, because they would have taken such a decision as evidence that we had lost our resolution. Resolution, an important quality in war, is normally defined as willingness to accept casualties in pursuit of victory, but it is also willingness to inflict casualties on the enemy. A demonstration would have wasted one of the two bombs the evidence makes clear were the minimum necessary to persuade the Japanese to abandon the war. It would certainly also have strengthened their belief that they could bloody us so badly in an attempted invasion that we would accede to more favorable terms than the unconditional surrender we insisted upon in the Potsdam Declaration.

Mitchell Wilson, a physicist who worked with Fermi at Columbia, favored a technical demonstration during the war. In his 1972 book *Passion to Know* he reports what he found of wartime attitudes when he visited Japan in the 1960s. "No Japanese scientist I spoke to," he concludes, "could think of a token demonstration that would have impressed the wartime Japanese political leaders." I think it's important that Wilson and I, coming as we did from opposite points of view about the value of a technical demonstration, arrived at identical convictions of its futility.

My next concern is the moral question. Were we wrong to use atomic weapons against Japan?

I often think by analogy. An analogy to the use of the atomic bomb that I find relevant emerges from my memory of grade school studies in San Francisco in the 1920s. We learned much about the San Francisco fire that followed the famous earthquake of 1906. The fire started because the earthquake damaged gas lines; it grew into a major conflagration because the earthquake also damaged water mains to such an extent that firemen couldn't use water to fight the

Luis W. Alvarez (LWA) and Walter C. Alvarez, 1912. (Photo: University of California, Lawrence Berkeley Laboratory)

Crew of the annual boat cruise, University of Chicago, early 1930s. (l to r:) Gordon Allen, Creighton Cunningham, Hugh Riddle, LWA. (Photo: University of California, Lawrence Berkeley Laboratory)

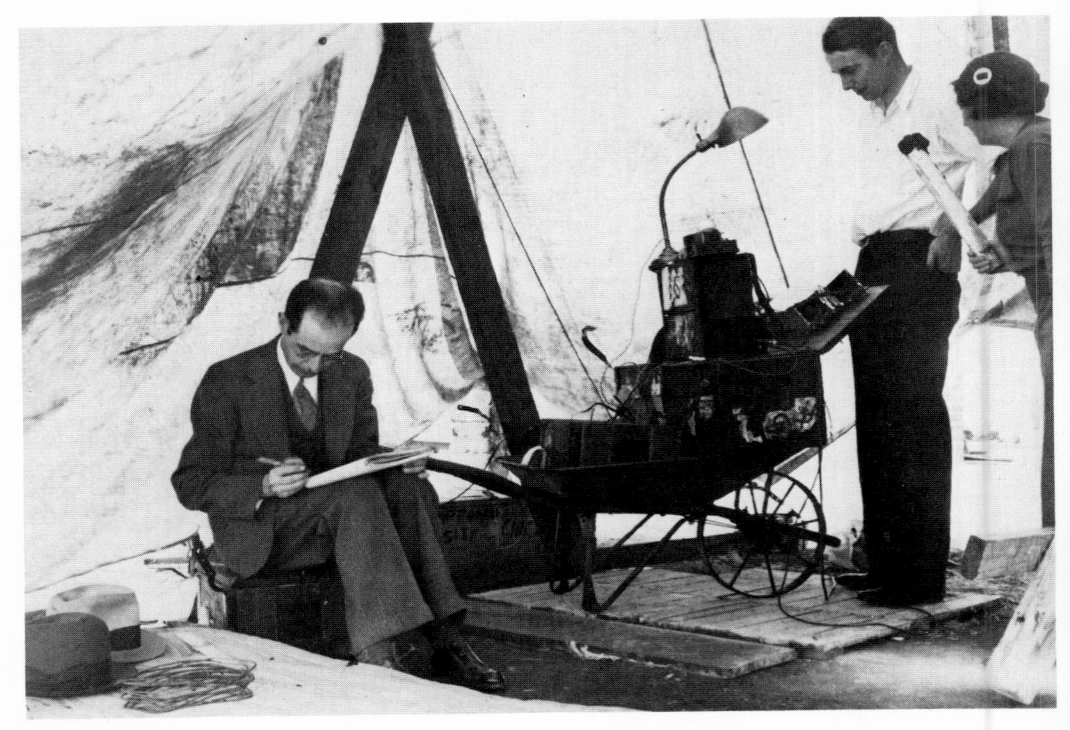

First good science: cosmic ray studies in Mexico City, 1933. (l to r:) Manuel Vallarta, LWA, Mrs. T. H. Johnson. (Photo: University of California, Lawrence Berkeley Laboratory)

Designated atmospheric scientist at the Century of Progress Exposition, Chicago, 1933, with T. J. O'Donnell. (Photo: University of California, Lawrence Berkeley Laboratory)

First job at Berkeley, designing the 60-inch cyclotron magnet, 1936. (Photo: Donald Cooksey, University of California, Lawrence Berkeley Laboratory)

Celebrating Ernest O. Lawrence's (EOL) Nobel Prize, 1939. (l to r:) Bob Cornog, EOL, Paul Aebersold, LWA. (Photo: Donald Cooksey)

Men who changed my life: (l to r:) Alfred Loomis, Sir Henry Tizard, Lee DuBridge, 1940, with a British magnetron. (Photo: The MIT Museum)

Last day in radar, MIT Radiation Laboratory, September, 1943. (Photo: University of California Lawrence Berkeley Laboratory)

On Tinian Island with parachute-borne pressure gauge and crew that monitored energy of Hiroshima and Nagasaki atomic explosions, August, 1945. (l to r:) Larry Johnston, Harold Agnew, LWA, Bernie Waldman. (Photo: University of California, Lawrence Berkeley Laboratory)

With Walter and B-29 model after the Japanese surrender, 1945. (Photo: University of California, Lawrence Berkeley Laboratory)

Receiving the Collier Trophy, White House, 1946. (Photo: University of California, Lawrence Berkeley Laboratory)

First operation of the 72-inch hydrogen bubble chamber, March 24, 1959. (l to r:) LWA, Don Gow, Bob Watt, Paul Hernandez. (Photo: University of California, Lawrence Berkeley Laboratory)

Celebrating news of the Nobel Prize, October 30, 1968. (Photo: Bill Young, University of California, Lawrence Berkeley Laboratory)

Jan and LWA dancing after the Nobel presentation, Stockholm, December 10, 1968. (Photo: University of California, Lawrence Berkeley Laboratory)

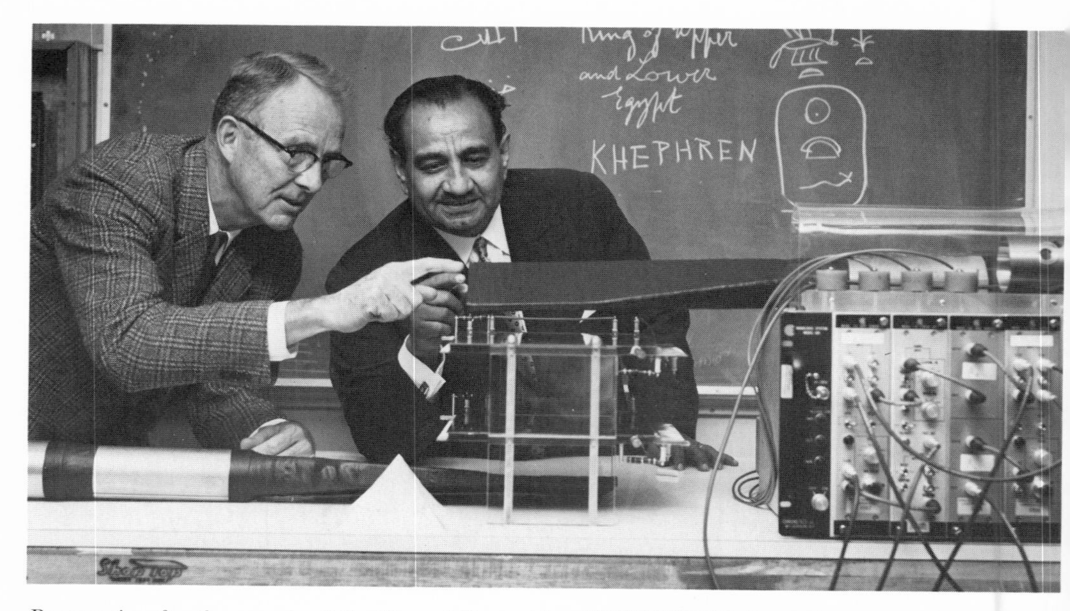

Preparation for the search of the Second Pyramid of Chephren for a burial chamber: LWA and Ahmed Fakhry with spark chamber, mid-1960s. (Photo: University of California, Lawrence Berkeley Laboratory)

Explaining the demise of the dinosaurs: LWA and Walter Alvarez at Gubbio K/T boundary, Italy, 1981. (Photo: University of California, Lawrence Berkeley Laboratory)

The Drs. Alvarez: three generations. (photo: Bernice A. Brownson, University of California, Lawrence Berkeley Laboratory)

fire. With no other recourse at hand, they began dynamiting houses in the fire's path to make firebreaks. Men who destroy other people's houses are usually considered criminals. But in April 1906 everyone understood that if those houses weren't dynamited they would soon be destroyed by fire but that, by dynamiting them, the fire fighters could save the houses beyond. Indeed, parties of civilians were actively constructing firebreaks in Hiroshima on the morning of August 6, 1945.

To me the crews of the *Enola Gay* and *Bock's Car* correspond to the San Francisco fire fighters. The hundreds of B–29's on the four airfields in the Marianas correspond to the approaching fire, which would have destroyed far more lives in Japan in the weeks to come by continuing to fly incendiary missions than were sacrificed in the creation of a firebreak. The atomic bombs created that break and stopped the war, the way firebreaks stopped the San Francisco fire. (I do not here mention the more terrible looming holocaust of invasion that we and the Japanese were spared, nor the U.S. Navy's proposals to starve the Japanese into submission by blockade. How many ways are there to die?)

If the Trinity shot had proved implosion unworkable, if the *Indianapolis* had been sunk three days earlier with the Little Boy bullet aboard, if the United States had for any reason decided not to use the atomic bomb in combat, Hiroshima and Nagasaki would certainly still have been destroyed, and very quickly. Like the emperor's palace in Tokyo and the temple city of Kyoto, the cities chosen for atomic bombing had been exempted from firebombing—in their case, so that the effects of the new weapons could be measured unambiguously (Nagasaki had been damaged by previous conventional bombing because it was not chosen as a 509th target until the end of July). I have never been able to understand why those who express revulsion at the use of the atomic bombs that ended a long and terrible slaughter so seldom question, or even bother to inform themselves on, the massive firebombing of Japan.

Did we need a second bomb? The historical evidence reveals no hesitation on the part of the Japanese military to continue the war. They had dared to attack and sink the U.S. Pacific fleet at Pearl Harbor, challenging a far more powerful nation than their own to retaliate; they could hardly be classed as fainthearted. They lost many times as many civilians in the course of the war as died at Hiroshima—a hundred thousand in one night during the firebombing of Tokyo on March 9, 1945, with a million casualties, several hundred thousand of them severe. In the Japanese tradition, surrender was unthinkable, and I find no evidence that military leaders even considered it after Hiroshima. With Nagasaki it was otherwise. The military leaders of Japan still could not agree to abandon their solemn oath to protect the person of the emperor by accepting unconditional surrender. Hirohito took the extraordinary step of forcing the issue. In his unprecedented radio broadcast to his people on August 15, he specifically mentioned "a new and most cruel bomb, the power of which to do damage is indeed incalculable. . . . This is the reason why We have ordered the acceptance of the provisions of the Joint Declaration of the Powers." Could anyone be in a better position to know why he decided to surrender than the emperor of Japan?

There is, finally, the question implied by the presence of all those waiting coffins I heard about on Tinian. What would Harry Truman have told the nation in 1946 if we had invaded the Japanese home islands and defeated their tenacious, dedicated people and sustained most probably some hundreds of thousands of American deaths and more hundreds of thousands of casualties and if the *New York Times* had broken the story of a stockpile of powerful secret weapons that cost two billion dollars to build but was not used, for whatever reasons of strategy or morality? The *Times* would have reported correctly that the President had controlled the power to stop the war in early August of 1945 with no further loss of life on either side. How would Truman have explained his decision at his impeachment trial? What would he have said to the parents and spouses and children of those who were killed? Indeed, what would he have said to the Japanese still alive on their burned and broken land?

Was the problem the Allied demand for unconditional surrender, as many seem to think? Foreign Minister Shigenori Togo had cabled the Japanese ambassador in Moscow on July 12, seeking Soviet mediation,

> It is His Majesty's heart's desire to see the swift termination of the war. . . . However, as long as America and England insist on unconditional surrender our country has no alternative but to see it through in an all-out effort for the sake of survival and the honor of the homeland.

I find no willingness to surrender in this message, but rather a Japanese hope by prolonged negotiations to persuade the two Allies to abandon a long-maintained demand. To the last the Japanese were determined to insist on the continued authority of the emperor. That was not what unconditional surrender meant to the American people, as Harry Truman and his secretary of state, Jimmy Byrnes, both skilled politicians, well knew. To most Americans in 1945, after four years of ugly war, Hirohito was a master war criminal who deserved to be hanged.

My young graduate students sometimes find it hard to believe me when I tell them that the people of the United States in 1945 were enthusiastically in favor of the bombing, including the atomic bombing, of Japan. I decided to demonstrate by going to the library and looking up the last issue of *Life* magazine before the Hiroshima story broke. *Life* was then for most Americans what network television news is today, their preeminent source of information about the world. Parents and children alike scanned it avidly; it was routinely used in schools. What I found in the last issue before Hiroshima was even worse than I had expected: a one-page spread of pictures entitled "A Jap Burns," eight photographs depicting a Japanese soldier being flushed from a shed with a flamethrower, his body and hair stuck with napalm and aflame, crashing through the jungle and collapsing to agonized death. "So long as the Jap refuses to come out of his holes and keeps killing," the text concludes, "this is the only way." Most Americans in 1945 enthusiastically agreed.

Qualifying our demand for unconditional surrender was politically impossible in such a setting. That isn't only my conclusion. The issue was joined directly when the Potsdam Declaration was being prepared in July. Secretary of War Henry Stimson proposed reserving the emperor's authority. Byrnes and Truman disagreed, precisely because they believed that such a step would look like appeasement to the American people and weakness to the Japanese. After Nagasaki, when the Japanese agreed to surrender on our terms, there was no objection from the American public to the softening of the terms, only great rejoicing. I can't invent another scenario in which the Gordian knot could have been cut so swiftly. Two powerful new bombs in quick succession shocked the Japanese into surrender. That was why we built them and why we dropped them.

Having said this much, I also have to express sorrow at the terrible loss of life on both sides in that last world war. I believe that the present stability of the world rests primarily on the existence of nuclear weapons, a Pandora's box I helped open with my tritium work and at Los Alamos. I believe nuclear weapons have persuaded the two superpowers to work together to defuse dangerous international conflicts. I'm happy that we apparently have broken the twenty-five-year cycle of European wars, two of which enlarged to world wars. I had something to do with this seminal change in the direction of world affairs. I feel great pride in that accomplishment.

I have difficulty understanding why so many people see nuclear weapons as mankind's greatest threat. Not one of them has been used since World War II, and without question they have prevented World War III, which would otherwise almost certainly have been fought by now with enormous loss of life.

I'm very much in favor of the eventual elimination of *all* weapons, both nuclear and conventional, though I doubt that anyone alive will see the day when that happens; it will take, I would guess, about a hundred years to come to so utopian a condition, time enough for all the old soldiers and their children and grandchildren to die off. By then, I think, we and the Russians will have achieved the same sense of stability and mutual trust that we have long enjoyed with Canada. The United States and Canada still maintain forts on their common border with cannons aimed across, but they're tourist attractions now. We came to mutual trust not by a series of arms control treaties and mutual inspection teams but simply by the passage of time, during which we learned to live in peace. I'm confident that we will learn similarly, given time—and time is what nuclear weapons have given us—to live in peace with the Soviets.

Although I don't count never changing one's mind as much of a virtue, I've held this view since I witnessed the destruction of Hiroshima. It may seem hawkish to some. I do agree that we have far more weapons than we need for secure deterrence. I would like to see that number reduced, but I would prefer that both sides keep some fraction of their nuclear arsenal, to ensure that we fight neither a nuclear nor a conventional war until the world has cooled down to the point where war itself can be safely abandoned.

The last few centuries have seen the world freed from several scourges—slavery, for example; death by torture for heretics; and, most recently, smallpox. I am optimistic enough to believe that the next scourge to disappear will be large-scale warfare—killed by the existence and nonuse of nuclear weapons.

NINE

Coming Back

I BEGAN THINKING about my postwar physics career before the war was over. It was evident that the future of nuclear physics lay in exploring those high-energy regions of which cosmic-ray phenomena gave a foretaste. To do so would require inventing newer and better particle accelerators. The cyclotron for protons and deuterons and the betatron for electrons were the two important prewar accelerator types. Each was limited in maximum energy—the cyclotron by the relativistic mass increase that caused particles to drop out of phase with the accelerating fields, the betatron by the intense electromagnetic radiation that resulted from the confinement of higher-energy electrons in circular orbits.

I became convinced that an electron linear accelerator (linac) was the best way to reach high energies. Bill Hansen at Stanford had first recognized that conducting resonant cavities (that is, evacuated metal cylinders to which tuned radio frequencies are applied) could produce high voltages. He lacked vacuum tubes that could supply sufficient RF power. I was one of Bill's most interested pupils in his weekly classes in the early days at MIT. A little before microwave radar came along, Bill showed Ernest Lawrence and me an evacuated copper cavity a meter in diameter and with a built-in triode. Bill believed he could achieve a million volts acceleration across the cavity's half-meter length. This first electron linac performed poorly because of problems in pumping the vacuum down and in insufficient RF power, but it contributed to the development of the klystron.

In the middle of the war new antiaircraft guns directed by microwave radar made the Army's 1.5-meter SCR–268 radar sets obsolete. Bureaucracies once in motion are hard to turn off, however, and by the end of the war the Army owned three thousand militarily worthless SCR–268's. Their high-powered, pulsed 1.5-meter oscillators were just what I needed to make a 2,000-foot-long 500-MeV electron linac. I calculated that 600 SCR–268 oscillators, each pulsing fifty kilowatts of power into an evacuated copper cavity, could provide electrons with more than enough energy to make muons, particles so far seen only in cosmic-ray collisions. I sent my thirty-page proposal for an electron accelerator to Ernest, who agreed to help me get the SCR–268's when the war was over.

Ed McMillan, then also at Los Alamos, was thinking about similar challenges. He came up with a spectacular invention that circumvented the cyclotron energy limit and raised the betatron limit as well. When Ed told me about it I shook his hand. His idea, called phase stability, requires in a linear accelerator that most of the particles under acceleration cross a gap before the electric field reaches its maximum value. Late particles thus experience a larger accelerating force, gain more velocity, and catch up; early particles speed up less and fall back into phase. Ed's design for a betatron-like phase-stable 300-MeV electron synchrotron looked so elegant, inexpensive, and obviously workable that I couldn't imagine my long accelerator competing.

But it wasn't clear to me that his concept would work as well for heavy particles; overnight I changed my linac plans from electrons to protons. Ernest happened to call me in the midst of these changes. I told him I now wanted to accelerate protons because Ed had just put me out of business. He wanted to know more. Not thinking, I outlined Ed's idea. Ernest immediately grasped its importance and became excited. After the call, realizing my error, I found Ed and apologized for denying him the pleasure of being the first to tell Ernest of his brilliant invention.

Ernest had also been thinking about the postwar period. After the calutron plant at Oak Ridge reached full production, early in 1945, he persuaded General Groves to authorize the conversion of the Berkeley Radiation Laboratory back to its peacetime mission of fundamental physical research. The Manhattan Engineer District (MED) accordingly made funds available to complete the 184-inch cyclotron and to build Ed's electron synchrotron and my proton linac.

Ed and I, who had both left Berkeley as assistant professors, both returned as full professors. Like Glenn Seaborg and others coming home from wartime assignments, we had gone away as boys, so to speak, and came back as men. We had initiated large technical projects and carried them to completion as directors of teams of scientists and technicians. We were prepared to reassume our subordinate roles with Ernest as our leader once again, but by his actions, if not in so many words, he signaled that we were to be free agents. We made all our own technical and personnel decisions, and for the first few years after the war we had unlimited financial backing from the MED. Although Ernest showed a keen interest in what we were doing, in those early postwar years he never gave a sign that he considered it his function to advise us in any way. Wise parents let their children solve all the problems they can but stand by to help when a problem proves too difficult. Ernest was always a wise scientific parent; all of us who were fortunate enough to be his scientific children remember with gratitude the help and understanding he offered when we needed it as well as the freedom he gave us to solve our own problems when it seemed that we could eventually succeed.

Designing a proton accelerator was difficult and had never been done. The accelerator geometry and electrical properties were completely different from those of an accelerator for electrons. I didn't figure out how I was going to accelerate protons until after I got back to Berkeley. Gerry, Walt, Jean, and I moved temporarily into Robert and Bernice Brode's redwood house in the Berkeley hills. Comfortable, well-furnished, with a beautiful view of the bay, the house was a pleasure after our dreary apartment at Los Alamos. In a few months we found a spacious Spanish-style house closer to the university. Even though I didn't know what my accelerator would look like inside, I pushed ahead full-blast assembling my team. The competition for the best young physicists was intense and the recruiting hectic. My first choice was Pief Panofsky, the brightest of the new crop I met during the war. Frank Oppenheimer, Robert's younger brother, a Caltech Ph.D., signed on. So did Larry Marshall, who had worked at MIT. Larry Johnston and Bruce Cork, my special wartime protégés, came back as graduate students and worked hard and effectively designing and building the linac.

We would assemble our accelerator in Building 10, the former calutron shop, in the middle of its three seventy-foot bays between a noisy welding shop and our shop-test area. I calculated that protons for injection into our linac would need 4 MeV of energy at the outset. Besides pioneering an entirely new kind of RF accelerator, then, our little group also had to work its way out onto the frontiers of direct-current high-voltage engineering, a field in which none of us had experience. We would get the 4 MeV from an electrostatic generator, a Van de Graaff; the most powerful such machine in the world at the time was the 3-MV pressure Van de Graaff at Los Alamos, built by Ray Herb of the University of Wisconsin.

Our proton linac would accelerate protons by pushing and pulling them through an electric field formed inside an evacuated resonant cavity. The RF field would oscillate—reverse itself—so the particles needed to be shielded from its effect during most of each cycle. I soon realized that I could shield the protons inside a series of hollow metal tubes—"drift tubes"—suspended inside the cavity, shorter at the injection end and increasing in length down the cavity to compensate for the particles' increasing acceleration. This arrangement, which saved power, came to be called Alvarez geometry. (I don't call it that; like Arthur Compton when discussing the Compton effect and Enrico Fermi when discussing the several Fermi effects and Fermi theories, I resort to circumlocution to avoid the immodesty of naming myself in physics talk.)

Bill Hansen thought my drift tube design encouraging. The first Wideröe linac, which Ernest Lawrence and Dave Sloan built at Berkeley before the war, had dozens of drift tubes. Mine bore a superficial resemblance to theirs but behaved completely differently; particles spent most of one electrical cycle in my drift tubes, in contrast to most of one-half cycle in the Wideröe design.

After a few months our design team had set the accelerator's basic parameters. The length of Building 10's central bay determined the tank's length, which in turn determined the total energy to which the machine could accelerate protons, 32 MeV—28 MeV added to the protons injected at 4 MeV. The SCR–268 oscillators piled up in an Oakland Port Terminal warehouse, eventually totaling some two thousand. I never visited the warehouse, and we used few of the surplus units; my radio engineers decided they could build better oscillators on their own.

The RF team worked up a series of copper cavities one meter in diameter and half a meter long to explore how to feed RF power into such devices. These cavities weren't protected by a vacuum, and when a large voltage was established across them they often discharged with a loud bang; people rushed to see the spectacular spark showers.

The accelerator needed rigidity to stay tuned electrically when it was evacuated and allow the protons to pass unimpeded down its length. We built a cylindrical vacuum tank forty feet long, slit like a hot-dog bun and hinged so that the upper half could be pivoted up and away. The electrical cavity inside, nearly as long as the outer tank, was supported on three pads. With this architecture the vacuum chamber could deform and distort but leave the cavity inside unstressed. The cavity looked like an inside-out airplane, a long tube with aluminum forming rings around its exterior spaced a foot apart and connected by long copper strips. Douglas Aircraft built it.

The Radiation Laboratory shops were remarkable in those days. For purposes of focusing I had specified that a thin beryllium foil should cover each drift tube entrance. Haydon Gordon, our chief mechanical designer, devised a complicated holder that screwed into place without twisting the fragile foil. He labeled his drawing of this mechanism, half a dozen parts with screw threads and pins, "Beryllium Foil Holder." Somewhere on the drawing he specified copper construction, but the machinists missed the specification; a few weeks later, no questions asked, they presented us with a beryllium foil holder machined out of rare, difficult-to-work beryllium. I doubt if any other shop in the world would have used beryllium without first calling up and checking.

The work moved fast. The cavity arrived from Douglas in the fall of 1946, and we installed it in the vacuum tank. We started learning how to feed it pulsed RF power. We installed the drift tubes on supporting copper stems, through which cooling water flowed. Pief Panofsky, a better mathematician than any of us, solved the tricky problem of making the RF magnetic field constant all along the forty-foot length of the accelerator.

Whenever we removed the relatively flimsy cavity from the vacuum tank, we supported it from a ten-inch-diameter iron tube that in turn was suspended by its ends from two electrically operated hoists. A long-faced Robert Oppenheimer stopped me one day on my way to my office in LeConte Hall to deliver a message from his brother: one of the hoists had malfunctioned. It had hoisted one end of the iron support; because there was no cutoff switch at the top, the chain separated and dropped the tube onto the cavity, completely crushing one end. When I saw what had happened to our carefully designed cavity, I was devastated. Fortunately no one was injured.

The accident benefited the project. We designed a new cavity that was better than the old one. Douglas Aircraft went back to work. With the good half of the old cavity, salvaged from the wreckage, we continued to learn about feeding RF power.

The new cavity arrived; we installed its drift tubes, adjusted its magnetic field, connected its high-power oscillators. We closed the tank, evacuated it, and applied power. When we opened the tank again, all the beryllium foils had disappeared. We were in serious trouble. Everyone had agreed that we couldn't make the accelerator work without the foils. They had simply burned out in the high electric fields between the drift tubes. Fortunately our mood was "Damn the torpedoes, full speed ahead." We looked for another solution to the focusing problem and found it, grids of tungsten strips mounted edge to beam and welded into the drift tube holes. Dashing off into the scientific unknown isn't for the fainthearted. You must believe you can find a way around any obstacles nature sets in your path.

We were ready then to marry the Van de Graaff generator to the accelerator tank. The protons had the right energy to match the first few drift tube spaces, but we couldn't get the beam very far down the line. I decided one evening that I understood the problem and assembled the team at the blackboard. With diagrams and equations I explained why the accelerator couldn't work in its present configuration and described a major change we would have to make. We found some minor changes we could make instead and fiddled with the system until midnight, when the beam emerged at its 32-MeV design energy, the highest-energy protons ever accelerated in a laboratory up to that time. We pounded each other on the back and celebrated. Someone later circulated a picture of my blackboard lecture. "9 P.M.," it was captioned: "Luie explains why accelerator can't work. First beam obtained at midnight the same evening."

Our proton linac was soon doing physics, particularly studies of proton-proton scattering. I used the beam to induce radioactivities in light nuclei and discovered several, including nitrogen 12, an isotope with a fleeting, one-hundredth of a second half-life that emitted high-energy positrons, and boron 8, a delayed alpha emitter important in the study of neutrino emission by the sun.

In mid-1946 Ernest told Ed McMillan and me that he was nominating us for election to the National Academy of Sciences. He was too busy to look up the necessary information, he said, and wanted each of us to write the other's letter of nomination. Ed and I both wrote good letters; in April the academy notified us of our election. (I modified this system for the many recommendations I came to write, asking applicants to draft their own letters and then rephrasing them. Such letters ought to be written by the person who knows the applicant's work best, and that is, of course, the applicant himself.) With one obvious exception, no award gave me more pleasure than election to the academy. My satisfaction came from having earned the approval of my peers in what I soon learned was unfettered judgment; members of the academy do not respond to pressure to elect someone they don't think worthy.

The first of many consulting jobs came along after the war. The Gilfillan Company of Los Angeles had made millions during the war manufacturing GCA radar sets; I signed on as a Gilfillan consultant at $200 a month, nicely augmenting my $6,500 Berkeley salary. I got a check every month for a dozen years until ITT bought the company and decided I was a luxury they needn't afford. GCA came back into prominence in the summer of 1948 when the Russians blockaded Berlin. The United States supplied the blockaded city by air for eleven months, and GCA at Tempelhof landed the steady stream of cargo planes through months of thick winter fog. With no other instrument-landing equipment available, the choice was GCA or surrender. We and they held out. (Eventually the instrument landing system—ILS—replaced GCA, except in the military, where GCA is still the favored system. ILS is less expensive because it doesn't require operators, and it's inherently less complicated. It works beautifully, as the hundreds of approaches I've made in my own plane attest. GCA had the field to itself during the war, however, and until after the Berlin blockade was lifted.)

The newly formed Rand Project of the Douglas Aircraft Company offered my second consulting job. Frank Collbohm persuaded Donald Douglas and Air Force Commanding General Hap Arnold that an independent group of specialists could help set Air Force policy for the future. Rand came into being with Frank as its director, and he immediately hired three consultants whom he had known at the MIT Radiation Laboratory: Dave Griggs, Louie Ridenour, and me. Our first assignment was to examine the possibility of orbiting artificial satellites and to think about their uses. That a satellite could be launched into low earth orbit was not then common knowledge. I traveled frequently to Santa Monica and persuaded Frank to hire Bob Serber as his first theoretical consultant.

Ed McMillan's theory of phase stability had indicated that the 184-inch cyclotron would perform better as a synchrocyclotron, which varies the frequency of the accelerating RF field to keep it in step with the particles as their relativistic mass increase reduces their frequency of revolution. No one had ever built a synchrocyclotron, and formidable problems were foreseen. Ernest and Ed had called for the immediate rebuilding of the old 37-inch cyclotron in the fall of 1945. It was soon operating as the world's first synchrocyclotron, and it demonstrated that such a mechanism was much simpler to build and operate than the 184-inch cyclotron originally proposed. By 1947, as a result of these early tests, the 184-inch machine was accelerating deuterons to 180 MeV and helium nuclei to 360 MeV, far beyond our 32-MeV protons—which still held the proton record, but the handwriting was on the wall. "By the way," Ernest said to me in the hall one day, "did I tell you that I've given your accelerator to the University of Southern California?" That was fine with me; it had served its purpose, and Ernest had properly concluded that it had no direct high-energy future even though it is the injector of choice into all high-energy proton synchrotrons.

With the completion of the 184-inch synchrocyclotron, Ernest once again became active in research. He had not participated directly in an experiment since his 1935 work on deuteron-induced radioactivities. After that work, until the war and the isotope separation project intervened, he had raised money so that the rest of us could do physics. Now, with adequate funding and fine administrators, he found time for personal participation again. Many of his young colleagues had known him largely as a laboratory director and as someone skillful at diagnosing troubles in complicated scientific tools. It was a refreshing experience for us to see the complete devotion with which he returned to basic scientific research.

The first discovery Ernest made was of a four-second delayed-neutron activity of low atomic number. The only known delayed-neutron emitters up to then were fission products. When the deuteron beam was first extracted from the 184-inch synchrocyclotron, it was dumped into a concrete block to keep it from scattering through the building. While using a hand-held neutron counter, Ernest noticed that the concrete was radioactive. After he determined the unknown radioactivity's half-life and neutron energy, he challenged us to identify it. I had the pleasure of showing that it was due to nitrogen 17.

Ernest then became convinced that the 184-inch machine could produce artificially the newly discovered pi mesons (pions), charged particles 273 times as massive as electrons (but both positively and negatively charged) and the carriers of the nuclear force, the existence of which had been predicted by the Japanese theorist Hideki Yukawa. Since the cyclotron made only 200-MeV deuterons and 400-MeV helium ions—individual nucleon energies, that is, of 100 MeV (deuterium = 2 nucleons, helium = 4 nucleons)—it shouldn't have been able to produce a 140-MeV pion mass. But Edward Teller and his student W. G. McMillan had made the unexpected prediction that the Heisenberg uncertainty principle as it applied to the nucleus should allow the production of pions under certain conditions. Ernest worked closely with Eugene Gardner designing experimental setups using nuclear emulsions as detectors, but their efforts proved unsuccessful.

In February 1948 C. M. G. Giulio Lattes, a young Brazilian physicist who was one of the pion's codiscoverers, arrived in Berkeley on a fellowship. In a careful analysis of Gene's experiments, Giulio quickly traced the problem to Gene's developing technique, which made his emulsions insensitive to pions. Gene resurrected his old apparatus, and within a few days the two men had managed several good exposures. They scanned them sitting side by side at their microscopes. Giulio brought to the task a familiarity with the tracks of pions shared by only a few physicists in the world at that time. His diligence was rewarded with success one evening about nine o'clock when he first observed the track of an artificially produced negative pion coming to rest in an exposed nuclear emulsion plate. He and Gene immediately called Ernest, tracing him to Trader Vic's in Oakland, where he was entertaining friends. Ernest brought his guests to the laboratory, looked through the microscope, and enjoyed one of the greatest thrills of his life.

For the next several weeks the 184-inch synchrocyclotron was devoted exclusively to producing and recording negatively charged pions. All of us scanned emulsion plates for these fascinating objects. As our procedures improved, and the yield of pions increased, the nuclear emulsions we scanned under extreme magnification showed a negative pion in every other field of view, far more populous than

C. F. Powell's Bristol group (with which Giulio had worked) had found in emulsions exposed to cosmic rays. (They were pleased to work with plates that showed one pion for every thousand fields of view.) Gene and Giulio modified their apparatus to observe positively charged pions, which they quickly found. Although Ernest had played a major role in both discoveries—designing the 184-inch cyclotron and the apparatus used in the experiment—he characteristically insisted that only Gene and Giulio should sign the historic paper reporting them. Theirs was the first paper in particle physics, a field where large accelerators create unstable particles by materializing energy.

In the summer of 1948 I attended an international physics meeting in Zurich, one of the first since the war to include German scientists as well as French and other nationalities. One of the principal speakers was the German theoretical physicist Werner Heisenberg. Those of us who had worked on the Manhattan Project held Heisenberg in low esteem because of some demonstrably untrue and self-serving statements he made at the end of the war. I didn't look forward to meeting him, even though I still admired him for his invention of quantum mechanics. We had a fruitful exchange of ideas, however, avoiding the points that would make our interaction unpleasant. Everyone was excited to hear about our artificially produced pions. I passed out pion-tracked emulsion plates as a salesman passes out business cards.

Before the conference I had stopped off in England. Lord Cherwell, Winston Churchill's wartime science adviser (Frederick Lindemann —Churchill had arranged his ennoblement early in the Second World War), had invited me to dine with him at Oxford. Since food was still rationed in England, I had brought canned hams for former Los Alamos friends. I was unprepared for the marvelous dinner Lord Cherwell served in his private quarters in college. His British colleagues, especially, were stunned by the thick steaks and flawless fruit. Cherwell, independently wealthy, never married, drank, smoked, or ate meat. While the rest of us gorged ourselves that night, he dined modestly on two soft-boiled eggs.

The pion discoveries rekindled Ernest's interest in looking for what we then called negatively charged protons. He was always a pure experimentalist, not unduly influenced by theoretical ideas. Everyone knew that the making of antiprotons would require six billion electron volts (BeV or GeV), out of reach of the relatively low-energy

184-inch cyclotron. But we now know of two negatively charged particles, the delta and the sigma, that are not much heavier than the proton. The sigma lifetime is long enough to have been detected by Ernest's technique, but its mass was just out of reach. The delta's mass was within reach, but its lifetime was too short. So only rather minor reasons prevented Ernest from detecting negatively charged particles with protonlike masses after all.

Ernest hoped that negatively charged particles with a protonlike mass might be produced by energetic protons much as negatively charged pions were, and, as I've said, he wasn't too far off the mark. In looking for these particles, however, he set his emulsion plates at a 90-degree angle to the target rather than the 180 degrees Gene and Giulio used. When the first exposure yielded a crowd of negative protonlike tracks, Ernest abandoned the regular cyclotron schedule and designed new experimental arrangements every day. The crew would open up the cyclotron tank, move lead blocks, reposition the target, relocate the emulsion plates, and pump the tank down for an early-evening exposure. Ernest was excited to think that he might have found the long-sought negative proton, which all of us—including Ernest—knew was impossible. Ed McMillan and I quickly and without comment joined together as a protective unit to make certain Ernest didn't announce his discovery prematurely to the press.

Every night for two weeks we waited with Ernest for the developed plates to emerge from the processing tanks so that we could help him scan them. We found so many negative protonlike tracks that Ernest in his excitement switched to X-ray film to shorten the development time. We suggested several experimental changes, but it was Ernest's experiment, so he called the shots.

To make sure these enigmatic particles came from the target, I had slots cut in a brass plate and positioned it one evening between the target and the film. At three the next morning the developed film revealed that the tracks came not from the target but from the vacuum chamber between the cyclotron dees. The particles appeared to be projected nearly backward by comparison with the ordinary protons that spiraled out of the ion source toward the target. One of us, probably Ed, then realized that the cyclotron ion source must not only be making positive hydrogen ions by stripping electrons but also be adding electrons to make negative hydrogen ions—protons with double electrons. When they were accelerated, these negative ions traveled in the opposite direction from that of the positive ions. Occasionally they collided with some of the few atoms of air gases remaining in the tank. Such a collision might strip away the extra electron, leaving a medium-energy hydrogen atom. Electrically neutral and therefore unaffected by the cyclotron's magnetic field, the hydrogen atom would fly off in a straight line like mud from a spinning tire. When it hit the film emulsion, it would lose its remaining electron and produce the "negative proton" track we observed.

The three of us said good night and went home. Ernest must have been terribly disappointed. Ed and I felt we had at least done a good job of protecting our longtime friend and mentor.

Plane trips east filled my postwar years. I came to have a more than casual acquaintance with all the commissioners of the Atomic Energy Commission and served on a number of committees.

MIT's Jerrold Zacharias organized the Hartwell Project to think about antisubmarine warfare. I enjoyed working again with friends from wartime radar work—Zach, Ed Purcell, Jerry Wiesner. We visited several submarine bases and what we saw demonstrated dramatically the sad state of our antisubmarine warfare systems.

On a visit to Key West we rode in a submarine. The base commander welcomed us as exceptional patriots; it was an extremely hot day in July, and he had found, he said, that the number of Washington visitors varied as the inverse fifth power of the Washington temperature. I had toured a captured German U-boat in San Francisco as a child and found the Navy submarine's officers' quarters spacious in contrast. The Navy had apparently benefited from the Pullman Company's experience in making small spaces habitable. We were given a demonstration of the analog computers that compute and set the direction of the sub's torpedoes. We picked up a tanker several miles away and stalked it at periscope depth. With the aiming problem entered into the sub's computer, we verified through the periscope what was happening and how well the computer updated its solution. Just when the problem was solved and the computer was about to simulate the release of torpedoes, we heard the propellers of a ship bearing down on us. To avoid our being rammed, the skipper ordered us to dive without changing course. We went down a hundred feet, returning to periscope depth when the sound of the intruding propellers faded. The targeting solution was still good.

Eventually we recommended that our submarines use passive sonar, as opposed to active pinging sonar, which commanders were wary of turning on in battle for obvious reasons. As a result the Navy designed and built offshore passive-listening arrays with much more extensive coverage than any system that could be carried aboard ship. Decades later I still saw the great impact of our report.

The 1953 Robertson Unidentified Flying Objects Panel was another committee service of note. UFOs had commanded press attention to the point where Washington was beginning to take them seriously. President Dwight Eisenhower asked the CIA to convene a panel to determine if UFOs were a potential security threat. We listened for a week while people who claimed to have seen UFOs retailed some of the wildest stories I'd ever heard. We found no threat, no phenomena attributable to foreign artifacts, and no need to revise current scientific concepts. The check that reimbursed me for travel expenses was drawn on the Pittsburgh bank account of a private citizen. Why this total stranger would send me a check puzzled me; I had come to associate the UFO panel with the Air Force because we met in the Air Force section of the Pentagon. I finally remembered we had been working for the CIA.

In 1949 Ernest and I saw a color television system demonstrated. We had never heard it suggested that TV pictures might be shown in color; I brought along a small permanent magnet to confirm that the display in fact involved electrons hitting a phosphor screen. The quality of the system we saw was impractically poor, but it started Ernest thinking about high-quality color television. Not long afterward he announced that he had invented a cathode-ray tube upon which color pictures could be shown. Ed McMillan and I weren't impressed with Ernest's primitive and unworkable color tube. He felt otherwise and without much difficulty secured extensive financial backing from Paramount Pictures. I had introduced Ernest to my friend Rowan Gaither, an attorney and administrator whom I had known since the war, who had engineered the transformation of the Rand Project into the Rand Corporation, and who became the first chairman of the Ford Foundation. Ernest and Rowan formed a partnership to develop and produce color tubes. They hired Ed and me as consultants, not to invent but to act as sounding boards for Ernest's ideas.

He then started trying them out. The Chromatic Television Company began in the garage of Ernest's Diablo Country Club summer home; he spent months of evenings working there with technicians and other consultants from the Berkeley Radiation Laboratory. He invited Ed, his wife, Molly, and me to see the first pictures on his tube, bubbling over as he always did about anything that he liked. Ed and I were embarrassed by the tube's poor picture quality and its even poorer commercial potential. He dismissed our doubts with a wave of his hand. There were technical problems that had to be solved, he said, but the tube was certain to do the job.

Soon thereafter Ernest realized that his first tube was a disaster. Having committed himself, he was forced to go ahead somehow and invent a first-class tube. I've never seen this factor listed along with necessity as the mother of invention. Edward Teller told everyone for years that he was going to make a hydrogen bomb. His Super proved unworkable, but since he had committed himself he couldn't walk away. He and Stanislaw Ulam ultimately invented a workable thermonuclear device. Public commitment is often an essential driving force in invention.

The exceptional tube Ernest eventually developed used an array of fine vertical wires set a centimeter from the tube face to accomplish postdeflection focusing and color selection. His system was producing high-quality color images when RCA was still struggling to produce any color image at all. RCA fought hard for its complicated three-gun shadow mask tube and eventually won, technically, commercially, and with the Federal Communications Commission. Magnetic fields produced by water pipes and home electrical wiring made a rainbow of colors between dark picture areas in RCA's early shadow mask tubes. These colors could be tuned out but reappeared if the set was moved. The Chromatron, as Ernest called his tube, suffered from no such color fringing and generated a much brighter picture because it lost no electrons on a mask.

For a demonstration on the East Coast, Ernest and his staff silk-screened phosphor stripes on a thick Lucite plate and bolted the plate to a continuously pumped metal vacuum system. Engineers attending the demonstration at Paramount in New York City were incredulous. No one in the radio industry ever demonstrated a product at such a workbench stage. Chromatic moved to a commercial building in East Oakland and produced a number of properly sealed glass tubes that performed remarkably well. This operation, which spent money extravagantly, resembled a downtown branch of the Radiation Laboratory, from which most of the Chromatic technical people had come. Ernest led a technical meeting every Saturday morning at which Ed, Don Gow, and I talked back; few of the regular staff felt secure enough to argue with Ernest on his daily visits.

Technically fine as Ernest's invention was, Paramount's bad business judgment doomed it. Consortiums of U.S. television companies twice negotiated with Paramount for a license to manufacture tubes under the Lawrence patents. On both occasions the Paramount negotiators, whose expertise was confined to dealing with their streetwise but unsophisticated Hollywood peers, fell flat on their faces. Their failure forced the television industry to use the RCA shadow mask tube for decades, until Sony took away part of the market by replacing the round holes with vertical slots. After so many years of development the shadow mask pictures finally surpassed those I saw on Ernest's Chromatrons. I wonder what a similar effort expended on the Chromatron system would have yielded.

Chromatic's bankruptcy dealt Ernest a severe blow. He always gambled heavily on his ideas, betting his full scientific reputation on the success of his latest machine. In scientific work he felt fully in control and inevitably made the right decision to save the enterprise. Chromatic's potential success, on the other hand, depended first of all on Paramount's business acumen; the company lost millions. Predictably, it blamed Ernest for its failure. I'm convinced that the resulting stress hastened his early death at fifty-seven in 1958.

TEN

Oppenheimer and the Super

THE EXPLOSION of Joe I, the first Soviet atomic bomb, on August 29, 1949, changed my life as momentously as the Tizard mission had in 1940. I was surprised that the Russians had managed to build a bomb so quickly. Before the war few experimental nuclear-physics papers came out of the USSR, and their authors' names were seldom well known. Visiting the country in 1956 and talking with Russian scientists, I realized that the elaborate state honors some of them had received were most probably rewards for their superhuman efforts training scientists and technicians for nuclear-weapons work.

Espionage also contributed its share, saving many man-years of research. There are no secrets in nature, but the numbers that Klaus Fuchs and his counterparts provided the Russians certainly advanced the first Soviet nuclear explosion by several years.

President Truman announced the explosion on September 23. It worried me. I had enjoyed four years of basic research away from military development work. Weapons development in the United States had slowed compared with wartime levels, which was natural enough. But it looked to me as if a crisis was at hand, and I wondered whether I shouldn't contribute to that work again. I remembered that Robert Oppenheimer had recruited me for Los Alamos in the first place by emphasizing the importance of building the Super, Edward Teller's wartime design for a hydrogen bomb. So far as I knew the Super program had been neglected these past four years. It occurred to me that the Russians might be working hard on a Super of their own and might succeed in building that potentially vastly more destructive weapon before we did. The only practical defense against such an eventuality seemed to be to get there first. At the same time I found myself hoping that fundamental physical principles might bar such a weapon from the world.

On October 5 I discussed my concerns with Wendell Latimer, the chairman of the Berkeley chemistry department, and found that he had arrived independently at the same conclusions. The next day I talked to Ernest Lawrence. He took the project very seriously—he had just himself talked to Wendell. We telephoned Edward Teller, back at Los Alamos after an interlude at the University of Chicago, and after talking with him decided to visit Los Alamos to find out how work on the Super had progressed since the war. Ernest was scheduled to leave the next day for Washington; together we caught a plane that night for Albuquerque.

At Los Alamos we met with Edward, the theoretical physicist George Gamow, Los Alamos associate director John Manley, and the mathematician Stan Ulam. They saw a good chance for the Super if a sufficient volume of tritium could be made available. Tritium in some form was necessary, of course, to fuel the important deuterium-tritium reaction, the lowest-temperature and therefore most easily ignitable thermonuclear reaction. They were preparing elaborate calculations to examine the Super's hydrodynamics that would run on the first early-model digital computers.

Edward accompanied us on the short flight from Los Alamos back to Albuquerque, and we continued talking until bedtime. To make tritium by breeding it from lithium 6, we needed neutrons. The obvious source of those neutrons was a nuclear reactor. However, like other aspects of the U.S. atomic-energy program, the building of nuclear reactors had slowed in the years immediately after the war; given limited available reactor capacity, neutrons for tritium breeding would compete with neutrons for plutonium breeding and thus with the ongoing manufacture of atomic bombs. Edward brought up the heavy-water reactor as an easy way to generate more neutrons (heavy water used as a reactor moderator absorbs fewer neutrons than graphite plus ordinary water, leaving more available for tritium breeding). Almost all the reactors of the Atomic Energy Commission (AEC) were graphite moderated; the only heavy-water reactor at hand was the small one Wally Zinn had built at the Argonne in 1944. Ernest and I were looking for something to do to help the Super program along. We decided we would get going at once to start a project to build production-scale heavy-water reactors for the AEC. To pursue that effort, I went along with him to Washington.

We met with a number of AEC and Defense Department officials there and found enthusiasm for our heavy-water project. David Lilienthal, the chairman of the AEC, was less than lukewarm. I found his behavior somewhat shocking. He turned his chair around and looked out the window and resisted even bare discussion. He obviously didn't like the idea of thermonuclear weapons. We fared better with the other four AEC commissioners, who seemed to approve what we were setting out to do.

About a month before the Russian explosion, Congressman Carl Hinshaw had called me at the laboratory to discuss the present state of air navigation. We talked for a long time. He found my views different from the official Civil Aeronautics Administration views and asked me to write him a detailed letter. I prepared a thirty-five-page document for his personal use, and on arrival in Washington I called him to tell him I would like to deliver it in person. I mentioned that Ernest was with me. The congressman asked me to hold on and said he would immediately call me back. He soon did and said he had just spoken with Senator Brien McMahon, who wanted Ernest and me to join them for lunch in his chambers.

Brien McMahon was then chairman of the Senate and House Joint Committee on Atomic Energy. We briefed the senator and the congressman on our reactor project. They were very happy to see some action in the field of thermonuclear weapons; they both told us they thought we were doing the right thing. They said they would visit Berkeley within ten days and asked us to call them if anything held up our plans.

We had intended to fly to New York and from there to Ottawa, Canada, to visit the one large heavy-water reactor then operating in North America, at Chalk River, a natural-uranium reactor built with Manhattan Project cooperation beginning near the end of World War

II. In New York, finding we couldn't book seats to Ottawa, Ernest and I went up to Columbia to see I. I. Rabi. Rabi was worried about the Russian explosion, too, and liked our plans. "It's certainly good to see the first team back in," he told us. "You fellows have been playing with your cyclotrons for four years. It's time you got back to work."

At Berkeley we discussed our reactor project with Don Cooksey, Ed McMillan, Bob Serber, Glenn Seaborg, and other Radiation Lab leaders. They all signed on. Then and in later meetings with Dave Griggs, Bob Christy, and the AEC reactor division's Lawrence Hafstad, we explored what and where we might build.

During the war the Manhattan Project had spurred the building of four different kinds of reactors—the original Chicago reactor, CP–1; the air-cooled Oak Ridge reactor; the water-cooled, graphite-moderated Hanford production reactors; and the Argonne heavy-water reactor. Since the war no new reactor had been built in the United States, though the AEC files bulged with designs. It looked to us as if reactors weren't getting built because people wanted to design them to perfection rather than working their way up, a case of the best proving the enemy of the good. (A committee that had been set up to promote new reactor designs was jokingly referred to as the Pile Prevention Committee. Enrico Fermi, concerned about the conservative nature of the AEC program, joked that what the country needed most was a bomb that didn't blow up and a reactor that did.) We at the Radiation Laboratory were hardly reactor experts—I had only my limited experience with Fermi at the Argonne to draw on—but we had a demonstrated capacity for rapidly building large scientific systems. That was what we were offering the AEC, and it seemed to be interested.

I didn't want to build reactors. I disliked the idea of building reactors. But I felt the country needed them. Later in October, Ernest informed me I had been elected to carry through our program. He had spent the weekend looking over sites and favored one northeast of Berkeley on Suisun Bay. "He says I will be director of the Suisun Laboratory," I wrote in a small diary I kept during this period. "I am therefore going on almost full time as director of a nonexistent laboratory on an unauthorized program. Cleared my desk in the linac building and . . . moved."

For the next week we worked night and day to assemble design proposals and cost estimates, preparing for the presentation we would eventually have to make. That was to take place in Washington, and I traveled there again at the end of October. Informally I discussed our plans with some of the AEC commissioners and with the scientists and engineers of the General Advisory Committee (GAC) to the AEC, of which Robert Oppenheimer, by then moved to Princeton as director of the Institute for Advanced Study, was chairman. The morning of the scheduled GAC presentation, I stood inside the main entrance to the AEC building and watched my friends and any number of famous military men go upstairs to a closed GAC session. The meeting lasted for some time. I watched the participants come back down the stairs, and then Robert came along and invited me and Bob Serber to lunch.

We went to a small restaurant near the AEC building. That was the first time I heard Robert's views on the building of the hydrogen bomb. He told me he didn't think the United States should build it. The main reason he gave was that if we built a hydrogen bomb then the Russians would build a hydrogen bomb, whereas if we didn't build a hydrogen bomb then the Russians wouldn't build a hydrogen bomb. I thought this point of view odd and incomprehensible. I told Robert that he might find his argument reassuring but that I doubted if he would find many Americans who would accept it.

For the first time I realized that the reactor program we were planning to start was not one that the top scientist at the AEC wanted. My diary ends on this incident. I didn't present the heavy-water reactor program formally to the GAC. I returned to Berkeley and went back to doing physics.

But not for long. In 1948 Bill Brobeck had convinced us that a proton synchrotron could be built in the multibillion-electron-volt range. Ernest had immediately assumed responsibility for securing financial backing for this Bevatron from Congress and the AEC. The Brookhaven National Laboratory had recently been established on Long Island and was simultaneously asking for support for a similar accelerator. The AEC eventually authorized the 6.2-GeV Bevatron at Berkeley and the 3-GeV Cosmotron at Brookhaven. The Cosmotron came into operation before the Bevatron; the principle of external injection into a synchrotron was untried, and Ernest had uncharacteristically decided to build a quarter-scale model first because he didn't want what he assumed would be his last big machine to be his first failure. The model worked within nine months of the decision to build it, and Ernest authorized construction of the Bevatron to proceed. Before long the building was finished, and much of the magnet structure was in place. But in the meantime the Soviet atomic bomb and our abortive reactor project had intervened. Early in 1950 Ernest stopped the Bevatron project and turned the laboratory's attention to what appeared to be an impending shortage of fissionable material.

Canada's known sources of high-quality uranium were nearly mined out. Turmoil in the Congo led to fear that the rich Shinkolobwe mine there might fall into Communist-bloc hands. The United States had few known sources of high-quality ore. But there was a plentiful supply of thorium, element 91, in the world, and the bombarding of thorium with neutrons transmutes it to a less well known fissionable isotope, uranium 233. Herb York had shown that high-energy deuterons striking heavy-element targets gave large numbers of neutrons. Win Salisbury suggested that an intense neutron source could produce U-233 electronuclearly. He calculated that one ampere of 1-GeV deuterons would yield 10^{20} neutrons per second; if those neutrons were all absorbed in thorium, they would make three kilograms of U-233 per day. They could equally well make tritium for Edward Teller's hydrogen bomb program. Ernest got excited at the prospect and decided we should follow it through.

The so-called Materials Testing Accelerator (MTA) occupied most of my time for the next two years. It was not a happy time.

Ernest, Ed McMillan, and I first discussed what sort of an accelerator the MTA should be. The 184-inch synchrocyclotron could achieve the required energy, but it was limited in beam current. Because of the high current and energy requirements, we decided that the MTA should be a linear accelerator of my design and of gigantic proportions.

We faced substantial technical problems. We could not use grid focusing at the drift tube entrances, because the grids, in large numbers, would absorb too large a fraction of the deuteron beam; the accelerator was planned to be one thousand feet long with hundreds of drift tubes. The only other beam-focusing method then known for proton or deuteron linacs was by longitudinal, solenoid-generated magnetic fields, so-called weak focusing. For weak focusing to be used, the aperture in the drift tubes down which the deuterons would be accelerated had to be three feet in diameter. That was a frightening number; the two-inch drift tube aperture in my proton linac had required an overall vacuum-tank diameter of three feet. The MTA tank would have to be sixty feet in diameter. Ernest loved technical extrapolation, and we had fun drawing such an accelerator. (Ten years later Ernest, Ed, and I were issued a patent on the "electronuclear reactor." Pief Panofsky could equally have qualified as a coinventor, but by then he had gone to Stanford and, like the rest of us, had long since ceased to be interested in such machines.)

I went with Ernest to meet Gwin Follis, the chairman of the board of Standard Oil of California. Ernest explained in detail what we were doing. He needed the assistance, he said, of a large industrial organization with a good scientific infrastructure. After about an hour of discussion Follis fully committed his company to help. In the next few days he activated a wholly owned subsidiary that drew scientists and technicians from various other divisions. Alex Hildebrand, the son of the distinguished Berkeley chemistry professor Joel Hildebrand and a chemical engineer, became our liaison. Alex, Bill Brobeck, and I constituted the MTA design team; we set up offices in the newly erected Bevatron building. During this period the Bevatron magnet coils were wound with heavy cable; from our second-floor offices we could watch the work.

To find a home for the MTA prototype, Ernest asked Washington for a list of possible sites. Most were abandoned military bases. One at Livermore, California, a square mile in area, offered mile-long runways, barracks buildings, a gymnasium, and a huge indoor swimming pool. As soon as Ernest and I saw it, he said, "Well, Luie, this is it." Livermore was then a sleepy village. An arch spanning its main street proclaimed its most notable achievement, giving birth to a world-champion heavyweight boxer: "Home of Maxie Baer." It became a large and prosperous city with a single major industry, the Lawrence Livermore National Laboratory.

Building the MTA was a tour de force. The vacuum tank that served as the resonant cavity, a cylinder sixty feet in diameter and sixty feet long, contained at that time the largest volume ever to be highly evacuated. Drift tubes hung inside from long stems. Each weighed tons, and we put wooden bridges between them over which we could crawl. Oscillators operating at 12.5 megahertz fed the accelerator's continuously powered cavity (previous linacs operated in a pulsed mode). The ion source supplied one ampere of deuterons, which produced one-fourth an ampere of accelerated current, the largest beam ever seen in an electronuclear machine. We were working in absolutely unexplored territory. The average electric field in our prototype MTA was about the same as in our proton linac, but the MTA's cross-sectional area and stored energy were about four hundred times greater.

Every new accelerator discharges stored energy by sparking. In the process the rough spots on the interior surface of the tank that cause the sparking are either cleaned up or made worse. I had the job of cleaning the MTA tank. Night after night for weeks on end, like a test pilot learning to fly a new plane, I sat at the control table gradually running the drift tube voltage up to the point where the tank would spark. I could hear that fearsome thunder all the way out in the control room. We weren't sure the tank would ever clean up enough to hold the necessary voltage. Calling the signals in full view of my Radiation Lab colleagues and the Standard Oil engineers who built the machine, I felt notably exposed.

As the weeks went by, I began to sense the machine's performance improving. The time it would operate at a given voltage gradient gradually increased. When it first operated at full voltage for an hour without sparking, we convened a great celebration in the control room. We then turned on the ion source, and as the deuterons flowed from one end of the tank to the other, we cheered.

The thousand-foot production reactor was supposed to be built in Missouri, President Truman's home state. The AEC hired the Bechtel Corporation to make a detailed design study of this monster; the resulting blueprints took nearly two hours to page through. Fortunately, it was never built. Capitalism ended the uranium shortage: the government offered large bonuses to any prospectors who found high-quality ore bodies. The Colorado plateau proved to be far richer in uranium ore than anyone had guessed. The MTA lost its purpose, and the project collapsed.

With the production accelerator canceled, we no longer needed the prototype. I disengaged myself from the project and returned to Berkeley to resume my career in research. Ernest, fascinated with the idea of a high-power linac, continued to back its development. Strong focusing, invented in 1952, reduced the size of the drift tube holes. Ernest and Standard Oil then built an accelerator with strong-focusing magnets that was no longer than our original MTA but operated at four times the frequency. Chester Van Atta became Ernest's linac expert. I had lost all interest in these devices and was happy to see him take over my job. They came close to achieving a technological breakthrough, but the twenty-megahertz accelerator was eventually dismantled; the MTA building came to house the Lawrence Livermore National Laboratory's magnetic fusion work. (Much of that innovative project was canceled early in 1986 when the price of oil took a downward turn.)

The prototype MTA attempted to carry a technology beyond its reasonable limits. Techniques appropriate to smaller machines are usually inappropriate to large ones. Only a succession of improvements have made possible the exponential growth of particle accelerator energy. The simple DC accelerator that John Cockcroft and E. T. S. Walton pioneered at the Cavendish in the early 1930s hit its limit at 1 MeV; Ernest invented the cyclotron to pioneer the new range above that energy. The synchrocyclotron avoided the cyclotron's 50-MeV limit; it was limited in turn to a few hundred MeV. The proton synchrotron, of which the Bevatron is an example, proved practical up to several GeV. Had the Bevatron with its weak focusing been extrapolated to higher energies, the cost and weight of the iron for its magnets would have become prohibitive. But the invention of strong focusing made it possible to build proton synchrotrons with energies to hundreds of GeV. Fermilab, the two-kilometer-diameter ring at Batavia, Illinois, came on line with an energy of nearly 500 GeV, and the invention of superconducting magnets has pushed that energy up to 1 TeV inside the same tunnel. Another invention, colliding-beam storage rings, increases such an accelerator's effective center-of-mass energy more than fiftyfold. In my opinion, in which I seem to be one of a small minority, some new invention will be necessary to extend accelerator energies still higher. Remember the MTA!

I had drifted far from experimental physics. The terrible feeling of receding from my chosen work led me to strange behavior. Back at Berkeley but no longer an active player, I contrived a series of rationalizations to avoid having lunch with those who were. I had always kept up my end of physics conversations; now that I wasn't active I had little to contribute and didn't completely understand what I heard. It was a new experience. It was also uncomfortable. I found myself lunching in the company of the scanners, technicians who examined nuclear emulsions. They were friendly, attractive young people, the conversation was light, and I felt diverted. It never occurred to me that I was really hiding out from my physics friends, afraid of exposing my ignorance.

For the first time in fifteen years I wasn't in the front lines. Since the war the Berkeley physics department had hired not one nuclear physicist; the thrust of work at the Radiation Laboratory now was particle physics. In a series of elegant experiments using the 184-inch synchrocyclotron, Burt Moyer, Herb York, and their colleagues had discovered the neutral pion. Soon afterward on the 300-MeV electron synchrotron Pief Panofsky and his group had determined the neutral pion's spin, parity, and mass. Through the years of work on the MTA, I couldn't spare the time to apprentice myself to Pief to learn this new business. I've watched the scientific careers of most of my friends and acquaintances permanently derail at similar points in their lives.

I returned to teaching physics. I still understood nuclear physics well. I wasn't successfully keeping abreast of particle physics, a rapidly expanding field. If not for my professorship I would probably have spent the rest of my days as an accelerator physicist, where I had demonstrated some talent. But since that field became much more mathematical I would eventually have drifted far from the mainstream.

My output of physics papers dropped way off in this period, as would be expected. But I published one very short letter that had a great impact on nuclear physics and started a new industry with sales of more than one hundred million dollars. The letter described a machine that is now usually called a tandem Van de Graaff accelerator. Robert Van de Graaff was the chief engineer for the High Voltage Engineering Corporation; he did a fine job of designing a practical tandem, which Willard Bennett invented in 1938 and which I invented independently—and reported in my letter—in 1951. Unfortunately, Bennett failed to report his invention in the open physics literature and apparently never tried to build one. His patent was quite unknown to accelerator physicists; it was only after my publication alerted HVEC to start building tandems that Bennett called attention to his patent. That oversight was too bad for both the nuclear-physics community and for Bennett. He received patent royalties on only the first tandem built by HVEC. After that his patent expired, and HVEC could use it without paying royalties. He would have been a very rich man if he had called attention to his patent as soon as the war was over. (If he hadn't patented his invention, my patent would have been owned by the government and I would have received the standard one dollar.)

A conversation with Herb York, one of the best things that ever happened to me, led me to find my way back. In 1952 Ernest expressed concern that all U.S. nuclear-weapons design was concentrated in a single government laboratory. He had great respect for Los Alamos and its accomplishments, but his extensive experience as a scientific consultant had taught him the benefits of healthy competition. He urged the AEC and the Joint Committee on Atomic Energy to establish a second laboratory. He offered the Livermore MTA site as a suitable location and pledged his personal oversight of the project.

Ernest picked Herb York to administer the new laboratory, with Edward Teller as de facto director. During the first U.S. nuclear test with a thermonuclear component, the Castle shot, Herb had run an important diagnostic experiment. He found his involvement in the weapons program challenging and accepted Ernest's offer. He left behind two graduate students who needed looking after, Lynn Stevenson and Frank Crawford. They had taken my nuclear-physics course, and they agreed to my supervision. They were my first new graduate students in five years.

I began thinking about basic physics again. All my previous work transitions had been easy. Getting back into physics at the age of forty was a chore. Like baseball players, physicists do their best work when they're young. As they approach forty, their skills erode. Baseball players open bowling alleys or bars or become managers. Physicists become deans or lead teams of younger men.

I faced the fact that I was a has-been and that my press clippings wouldn't impress the young. To get back into physics, I would have to learn their language. I would have to convince them I could contribute if they let me work on their experiments. Frank and Lynn were versatile and talented, so I made a deal with them: I would hire them as my research assistants if they would treat me as *their* research assistant. Once I convinced them I was serious, they assigned me homework and let me help. The experience was hard on the ego but exciting. Our desks were pushed together with those of Don Gow and Hugh Bradner, who were also taking this impromptu refresher course. Ernest joked about my living in the bullpen but recognized what we were doing and wished he had the freedom to join us.

Frank, Lynn, and I watched for an experiment we could do together that related to their thesis work. At one weekly research progress meeting a theoretician commented that if pions materialized some of the time in nuclei, as everyone believed, then gamma-ray scattering from nuclei should be increased. That looked interesting. If one pion was present in a nucleus all the time, the proton-scattering cross section would be fifty times larger. If a pion materialized 20 percent of the time, the cross section would be ten times larger, a measurable effect but certainly not an easily measurable one; gamma rays would still scatter from the electrons in the hydrogen atoms a million times more frequently than from the protons.

We spent several weeks designing an experiment to discriminate between gamma-ray scattering by protons and by electrons. Eighty curies of fission products provided a copious supply of 1.6-MeV gamma rays. We scattered them from fifteen-kilogram rings of carbon and plastic material into a sodium iodide counter after screening out the background with a lead absorber over the counter and a one-foot-thick uranium brick between the source and the counter. With this and other gear we measured the scattering of gamma rays by protons for the first time. Our data agreed with a prediction made sixty years earlier by the English experimental physicist J. J. Thomson, the discoverer of the electron. We saw no signs of pions materializing in the proton. Our theoretical friend went back to his drawing board, found a mistake, and agreed that the effect should be unobservable.

Lynn, Frank, and I were happy working together. That was the most valuable result of our year of hard effort. The personal equation is important in science, though it receives little attention in the literature.

After a long hiatus I was back doing real physics. When I next visited the Brookhaven National Laboratory, I spent my time on the Cosmotron floor talking to the young physicists about their experiments and answering questions about ours. Two years before I would have hung around the director's office reminiscing about the good old days. That change was the proof I needed that I'd been rehabilitated.

Late in 1953 the Atomic Energy Commission proposed to suspend Robert Oppenheimer's security clearance and AEC employment, writing him, "There has developed considerable question whether your continued employment on Atomic Energy Commission work will endanger the common defense and security." Robert responded by requesting a hearing, and that well-known hearing "In the matter of J. Robert Oppenheimer," the transcript of which was subsequently published and is still in print today, began in April 1954. Government lawyers had traveled across the country interviewing those of us who participated in the secret debate on the desirability of inventing and building a thermonuclear weapon in response to the Soviet atomic bomb. They were taken with the diary I kept of that period and told me I would be asked to read it into the hearing record. They said they would like to question Ernest and me on the record, which would be confidential. Ernest assured them we would appear when summoned.

Shortly before that summons came, Ernest telephoned me from Oak Ridge, where he was attending an AEC laboratory directors' meeting, and announced emotionally that he wouldn't testify and that I shouldn't either. He said people had convinced him that the Radiation Laboratory would be greatly harmed if he testified and that he, Ken Pitzer, Wendell Latimer, and I were viewed as a cabal bent on destroying Robert. I had never seen Ernest intimidated before. He was suffering terribly from the ulcerative colitis that soon afterward claimed his life.

I canceled my plane reservations and informed the government lawyer who had scheduled our testimony that Ernest had decided we wouldn't be available as witnesses. The lawyer responded with disappointment but didn't ask me to change my mind. I had dreaded testifying and was pleased that I could forget the matter.

Lewis Strauss, the chairman of the AEC, called me at home that evening and challenged me. He said Ernest's unexpected collapse under pressure could be attributed to health problems but wondered what my excuse was for letting him down. He emphasized that I had important information that the hearing board shouldn't be denied. I worked for Ernest, I replied, and he had ordered me not to testify. I had a duty to serve my country, Lewis countered. I said I had served my country during the war. Lewis's emotional intensity increased as he ran out of arguments. As a parting shot he prophesied that if I didn't come to Washington the next day I wouldn't be able to look myself in the mirror for the rest of my life. I had never disobeyed Ernest's direct orders, I said, and I wasn't about to start now.

After I hung up I sat thinking about Lewis's challenge. I finally concluded that I really would be ashamed to think that I'd been intimidated. So I poured myself a stiff drink, booked a seat on the TWA midnight red-eye flight, and sent Lewis a telegram. Later that evening I drove to the airport to disobey Ernest for the first time and to give testimony that might hurt a friend, though I hoped it wouldn't.

In my testimony I told of my admiration and respect for Robert. I said I thought his reasoning in the H-bomb decision was faulty but in no way related to his loyalty to his country, of which I had no doubt.

All of us who were touched by the hearing were in some way wounded. No one has ever directly criticized me for testifying as a government, though not unfriendly, witness. I've felt disapproval but it hasn't been expressed, nor am I aware of having lost any friends. In the years after the hearing, Robert and I had a number of pleasant conversations at various conferences, about physics and the old days at Berkeley. Our long friendship was equal to the task of ignoring the hurts we had both experienced along the way to doing what we saw as our duty. But neither of us ever referred to the painful hearing days.

I am enormously proud of my part in the development of the atomic bomb and in the revitalization of the thermonuclear program. I feel strongly that the prevention of nuclear war is the single most important problem facing the world and enthusiastically support the long-held U.S. policy of deterrence by mutual assured destruction. It's certainly not a perfect system, but it has worked well for decades. I'm confident it will continue to work until it's no longer needed. No one has come up with a better system even though many of our best minds have been devoted for years to doing so.

ELEVEN

Hydrogen Bubble Chambers

MOST OF US who become experimental physicists do so for two reasons: we love the tools of physics for their intrinsic beauty, and we dream of finding new secrets of nature as important and exciting as those our scientific heroes revealed. But we walk a narrow path with pitfalls on either side. If we spend all our time developing equipment, we risk being called plumbers. If we merely use tools others have developed, we risk being censured by our peers as parasites. Through the later 1950s and early 1960s, the fifth decade of my life, I had the pleasure both of developing valuable new physics tools and of using them to make significant discoveries.

When I received my B.S. degree in 1932, only three of the fundamental particles of physics were known. Every bit of matter in the universe was thought to consist solely of protons and electrons; the massless photon was the quantum of light. In that year of my undergraduate degree the number of known particles suddenly nearly doubled when James Chadwick discovered the neutron and Carl Anderson the positron. Since then the list of known particles has lengthened rapidly but not steadily, its discoveries coming in bursts of active work.

The photon mediates the forces of electromagnetism. In 1935 the

Japanese theoretician Hideki Yukawa published an important paper asking what the characteristics might be of a particle that similarly mediated the forces of the nucleus. He predicted that such a particle should have a rest mass about two hundred times that of an electron, an energy at rest of about 100 MeV. In 1937 Seth Neddermeyer and Carl Anderson found a particle with that mass in cosmic rays, the "mesotron." Most nuclear physicists spent the war years secure in the belief that the mesotron was the particle Yukawa proposed and that it would be available for study when hostilities ceased.

As it turned out, they were wrong. Three young Italian physicists hiding from the Germans in a cellar in Rome originated modern particle physics by demonstrating that Neddermeyer and Anderson's muon, as it came to be called, was almost completely unreactive. If it mediated nuclear forces, it should have been highly reactive. The physics community fortunately had to endure this confusion for less than a year, from the Italian paper in 1946 until the cosmic-ray discovery in 1947 of the pion, a singly charged particle that fulfilled the Yukawa prediction of vigorous reactivity and decayed into the unreactive muon. Positive and negative pions turned up first; the neutral pion was discovered at Berkeley by Herbert York in 1950. Important experiments on the interactions between pions and nuclear particles were then done, notably by Enrico Fermi and Herb Anderson at the University of Chicago beginning in 1952.

Although the study of the production and interaction of pions had passed in a decisive way from the cosmic-ray groups to the accelerator laboratories in the late 1940s, the cosmic-ray-oriented physicists soon found two new families of particles characterized by a quantum-physical condition that came to be called strangeness. In contrast to the discovery of the pion, which almost everyone accepted immediately, the discovery and the eventual acceptance of strange particles stretched out over a period of several years. Such heavy, unstable particles were first seen in a cosmic-ray-triggered cloud chamber in 1947. By the time our Bevatron first operated, in 1954, a number of different strange particles had been identified: several charged particles and a neutral one all with masses in the neighborhood of 500 MeV; and three kinds of particles heavier than the proton.

Those who are familiar with particle physics will wonder why I said there were several charged particles with masses near 500 MeV when we now know of only one such particle, the K meson. My young colleagues and I had what in baseball would be called an "assist" in shrinking the several kinds of K mesons down to a single variety. It was conventional wisdom in 1955 that all of the K mesons couldn't be one particle; that would violate the law of the conservation of parity. T. D. Lee and Frank Yang agonized over this puzzle for months and finally suggested that parity violation might take place in "weak" decays.

Ernest Lawrence, a pure experimentalist, reacted to the famous theta-tau puzzle almost instantaneously. One night Ernest dropped in on Lynn Stevenson, Frank Crawford, and me as we were measuring the lifetimes of various kinds of K particles on the Bevatron floor. I was measuring photographs of oscilloscope traces and heard Ernest ask Lynn what he was doing. "We're trying to find a difference in the lifetimes of four different kinds of K particles," Lynn said, "and they're getting closer every day." Ernest was spending most of his time then organizing the Livermore weapons laboratory and inventing color television tubes, so he wasn't up to date on the niceties of K-meson physics. After Lynn had explained the puzzle in some detail, Ernest said, "You mean to tell me that all these particles have very nearly the same mass and the same half-lives and yet they aren't the same particle?" Lynn said the theorists had reasons why the particles couldn't be the same. Ernest said, "Don't you worry about it—the theorists will find a way to make them all the same." And that, of course, is what Lee and Yang finally did, but long after Ernest had disposed of the problem in his characteristically intuitive way.

Now that I've placed myself at the Bevatron as an electronic physicist rather than the bubble chamber physicist I soon became, I'll add an adventure of that period as described by my good friend Gerson Goldhaber. It's entitled "The Finger in the Dike Revisited"; Gerson recalled it at a recent conference on the history of physics:

> I remember in particular one episode when Luis Alvarez was also spending the evening at the Bevatron and observing our procedures. That night the Bevatron operator charged with pulling the target probe, with our emulsions on the end of it, out through the vacuum lock gave a particularly vigorous pull and managed to yank the probe completely out and air started rushing into the vacuum tank. Luie, who was standing nearby, rushed over and placed the palm of his hand over the hole! This allowed the crew to close the vacuum lock without the entire Bevatron coming up to air. I must admit that I would not have thought of doing this—and furthermore would probably not have done it! Luie had saved the day and the Bevatron was able to pump back down without excessive loss of time, while Luie was rubbing the sore spot on his hand.

The strange particles all had lifetimes shorter than those of any known particles except the neutral pion. They were called strange particles because their observed lifetimes puzzled theoreticians. The puzzle was not that they decayed so rapidly but rather that they lasted so much longer, almost a million million times longer, than could be explained.

We could produce them with the Bevatron and wanted to study them, but none of the existing techniques used to track particles was well suited to studying the basic reactions of strange-particle physics. The diffusion cloud chamber, filled with gas, registered too few interesting tracks. The older Wilson cloud chamber operated on too slow a cycle. Nuclear film emulsions offered no way to resolve occurrence times. We concluded that particle physicists needed a track-recording device with solid or liquid density (to increase the rate of production of nuclear events over gas-filled devices by a factor of several hundred) with uniform sensitivity (to record events throughout its volume) and with fast cycling time.

In April 1953, with these concerns in mind, I made my annual trip to Washington to attend the biggest meeting of the American Physical Society. After lunch on the first day, I found myself seated at a large table in the garden of the Shoreham Hotel. All the seats but one were occupied by old friends from World War II days, and we reminisced about our experiences at MIT and Los Alamos. A young man who had not experienced those exciting days sat next to me. We began talking physics. He said he was afraid no one would hear his contributed paper, because it was scheduled to be the last paper presented in the last session on the last day.

In those years of slow airplanes almost everyone left early. I admitted I wouldn't be on hand for his paper and asked him to tell me about it. He was a postdoc at the University of Michigan; his name was Don Glaser. He had invented a new particle detector, he said, which he called a bubble chamber. I had heard rumors of his device but knew nothing of its properties. He showed me photographs of bubble tracks in a small glass bulb about one centimeter in diameter and two centimeters long filled with ether. By suddenly reducing the pressure on the ether to a level below that at which ether usually boils—superheating—he had produced a temporary state during which the disturbance of a charged particle passing through the liquid caused it to boil only where the particle passed, leaving a visible track of small bubbles. He said that he could maintain the ether in a superheated state for an average of many seconds before spontaneous boiling took place.

I immediately realized that this new detection technique, which had condensed-matter density and was track forming, could be the basis of an ideal detector for the Bevatron. That night in my hotel room I told Frank Crawford and Don Gow what I had learned. We discussed building a big bubble chamber using not ether but liquid hydrogen, in which nuclear interactions could take place only on the simplest nucleus, the hydrogen atom's single proton. Frank volunteered to stop off in Michigan on the way back to Berkeley to learn everything possible about Don Glaser's technique.

I returned to Berkeley on Sunday, May 1, and the next day Lynn Stevenson began keeping a new laboratory notebook on bubble chambers. Frank got back a few days later. In a series of strategy meetings, we planned a development program for a liquid-hydrogen bubble chamber. We would first repeat Glaser's work by building a hydrocarbon bubble chamber and then move to liquid nitrogen and finally to the much more demanding and colder liquid hydrogen.

Lynn and Frank moved into the student shop in the electron synchrotron building. With the help of two talented synchrotron technicians, John Wood and A. J. ("Pete") Schwemin, they confirmed Glaser's discovery of radiation sensitivity in ether. Then they built a glass chamber installed in a Dewar flask—a Thermos bottle—to work down in temperature first to liquid nitrogen and then to liquid hydrogen. They found that although liquid nitrogen boiled more rapidly near a gamma-ray source, it didn't produce tracks.

They worked carefully down in temperature toward liquid hydrogen, aware of its nasty reputation as an explosive fire hazard. In the meantime John Wood, who had built liquid-hydrogen targets and wasn't afraid of the substance, designed and built a 1.5-inch glass liquid-hydrogen chamber. Pete Schwemin prepared an optical system with which to photograph any tracks that formed.

We were tremendously excited by John's liquid-hydrogen chamber. It was the first to show tracks forming in hydrogen. It also revealed a crucial difference between smaller and larger bubble chambers. John had never written an article for publication. I drafted one for him for the *Physical Review.* "We were discouraged by our inability to attain the long times of superheat," it noted, "until the track photographs showed that it was not important in the successful operation of a large bubble chamber." I have always felt that, second to

Don Glaser's discovery of track formation, John's was the key observation in the whole development of bubble chamber technique. As long as the chamber expanded rapidly enough, bubbles forming on the walls didn't destroy the superheated condition of the main volume of liquid; it remained sensitive as a track-recording medium.

Pete Schwemin got excited in turn and proceeded to build the first metal bubble chamber with glass windows. All earlier chambers had been constructed completely of smooth glass, without joints, to prevent accidental boiling initiated at the microscopic sharp peaks that project from rough spots. Pete's 2.5-inch was the world's first "dirty" chamber, a term I've never liked that came into use for a while to distinguish chambers with windows gasketed to metal bodies from all-glass chambers. Because it was "dirty," the 2.5-inch chamber boiled at the walls but still showed good tracks throughout its volume. "Clean" chambers soon became things of the past, and "dirty" chambers were no longer stigmatized.

Pete immediately started on a three-inch chamber. I suggested we maintain the exponential growth curve we'd started, so he moved up an increment and quickly built a four-inch chamber. That one went into a negative pion beam at the Bevatron. We had no stereoscopic cameras yet and no magnetic field, but we saw some particle collisions and some strange particles. Later the four-inch chamber was properly equipped and supplied millions of photographs of photon-electron interactions at the electron synchrotron.

A ten-inch chamber was to follow. In December 1954, shortly after the four-inch had been operated in the cyclotron building for the first time, I concluded that the ten-inch chamber we had just started to design would not be nearly large enough to tell us what we wanted to know about strange particles. The tracks of a few such objects had been photographed at Brookhaven, and we knew they were produced copiously by the Bevatron. We needed a bigger chamber to see them well.

Several different criteria set its size. Fortunately, all of them could be satisfied by one design; a designer of new equipment often finds that one essential criterion can be met only if the object is very large while an equally important criterion demands that it be very small. All dirty chambers built up to then had been cylindrical in shape and characterized by their diameter measurement. Studying the decay of strange particles, I convinced myself that the big chamber should be rectangular, with a length of at least 30 inches. To make sure there would be enough hydrogen upstream from the decay region where reactions would take place, I then increased that length to 50 inches. When I realized that the depth of the chamber could properly be less than its width, I changed the length to 72 inches without altering the volume.

The result of this straightforward analysis was a rather frightening set of numbers. The chamber was to be 72 inches long, 20 inches wide, and 15 inches deep. It had to be pervaded by a magnetic field of 15,000 gauss, so its magnet would weigh at least a hundred tons and would require two or three megawatts of power to energize it. It would require a glass window 75 inches by 23 inches by 5 inches to withstand the operating pressure of eight atmospheres, a force on the glass of a hundred tons. Only the first hydrogen bomb builders had any experience with such large volumes of liquid hydrogen; the hydrogen-oxygen engines of the Apollo lunar rockets were still gleams in the eyes of their designers—these were pre-Sputnik days. We were particularly worried about safety. Low-temperature laboratories were reputed to be dangerous places to work, even though they used much smaller quantities of liquid hydrogen than we were contemplating and maintained them at atmospheric pressure. I am deeply indebted in this regard to Bob Watt, who was largely responsible for the fact that we had no accidents with our hydrogen chambers.

Nor had anyone ever cast and polished such a large piece of optical glass. (The large telescopes—the 200-inch at Palomar, for example—are mirrors; their glass traps so many bubbles it's hardly even transparent.) Fortunately for the eventual success of the project, I persuaded myself that the chamber body could be constructed of a hollow plastic cylinder with metal end plates. My engineering colleagues later demolished this notion, but it kept the project alive in my mind until I was convinced that the glass window could be built. We weren't pushing the state of the art so much as sailing over the edge. I remember one day looking through a list of talks at a cryogenics conference and spotting one that read "Large glass window for viewing liquid hydrogen." I turned eagerly to the abstract only to find that the window it described, built into a Dewar flask, was one inch across.

In April 1955, after a few months of hard thinking, I wrote a prospectus entitled *The Bubble Chamber Program at UCRL.* It demonstrated in some detail why the large chamber was important to build and outlined a whole new way to do high-energy physics using such a device. It stressed the need for semiautomatic measuring devices (which had not previously been proposed) and described how electronic computers would reconstruct tracks in three-dimensional space, compute momenta, and solve problems in relativistic mechanics. All these techniques came to be standard. In 1955 no one had yet applied them. Of the many papers I've written in my life, I reread none with more satisfaction than this unpublished prospectus.

The big chamber, with building and power supplies, would cost $2.5 million (in 1955 dollars). It was clear that a special AEC appropriation would be required. I asked Ernest Lawrence if he would help me raise these funds. He read my prospectus. He then asked me to remind him of the size of the world's largest hydrogen bubble chamber. Four inches, I told him. He said he thought 72 inches was too large an extrapolation from 4 inches. I explained that the 10-inch chamber was on the drawing board; because we had designed the big chamber so that it could be considered hydraulically to be a collection of 10-inch chambers, if we could make the 10-inch work then we could make the 72-inch work. I thought it incongruous that Ernest, from whom I had learned so much about large-scale extrapolation, now felt I was being foolhardy. But he characteristically voted for the man despite his doubts about the program. "I don't believe in your big chamber," he told me, "but I do believe in you, so I'll help you get the money."

Ernest asked me to promise him that I would stick with the chamber until it was built and running. I do flit from one project to the next, although I like to think that I never left any of them, except the MTA, before its success was assured. Ernest clearly didn't recall my career in the same way. I regret deeply that he died a year before the 72-inch chamber became operational; he missed seeing that I kept my promise and that the chamber proved more successful than even I had imagined.

Ernest and I went to Washington and talked in one day to three of the five AEC commissioners: Chairman Lewis Strauss, the chemist Willard Libby, and Johnny von Neumann, the brilliant Hungarian-born mathematician. At a cocktail party at Johnny's home that evening, I learned that the commission had voted the same afternoon to award the laboratory the $2.5 million. All we had to do then was to build the chamber and make it work.

Design development had been under way for some time; now we speeded it up. Don Gow assumed a new role not common in physics laboratories: he became my chief of staff to coordinate the work of the physicists and engineers. He assumed full responsibility for the careful spending of our precious grant and undertook to become expert in all the technical phases of the operation, from low-temperature thermodynamics to safety engineering. He succeeded wonderfully but died young, a suicide; my younger son, Donald, is named in his memory.

I was deeply involved in the chamber design. At one point I deliberately designed myself into a corner, trusting that I could design my way out when the time came. All previous chambers had required two windows so that they could be illuminated from one side and photographed from the other. Such a configuration reduces the attainable magnetic field by excluding a lower pole piece, which would interfere with the light-projection system. I decided to design the 72-inch chamber with only one window. That would allow for a lower pole piece, increasing the magnetic field. It would also save the cost of a second glass window and improve safety by eliminating a potential weak point through which liquid hydrogen could spill.

The only difficulty with my decision was that for more than a year, while we firmed up the bubble chamber design and fabricated the mechanism's parts, none of us could invent a way both to illuminate and to photograph the bubble tracks through the same window. We tried dozens of schemes that didn't quite do the job. We did slowly come to understand our problems, and we eventually devised the retrodirecting system known as coat hangers, a series of odd-shaped reflectors that guided the light back precisely along the path on which it had arrived but that didn't reflect the bubbles. Our solution came none too soon; if it had been delayed another month, the initial operation of the 72-inch chamber would have been similarly postponed.

We took many other calculated risks in designing our system; if we had postponed fabricating the major hardware until we had solved all our problems on paper, the work might still not be in hand. Engineers are by nature conservative; they risk more than disgrace if a bridge collapses or a boiler explodes. Fortunately Paul Hernandez, our chief engineer, took even our most outlandish proposals seriously and applied all his ingenuity to making them safe before giving up those that proved irremediably flawed.

While we were building the 72-inch chamber, my colleagues Hugh Bradner and Jack Franck pursued improved measurement systems.

Cloud chamber physicists traditionally measured the angles and curvature of particle tracks on a space table, a flat white gimbal-mounted plane tediously rotated and moved so as to reproject stereo film into the region corresponding to the position where the tracks were photographed. Had we used such techniques, we could not have kept up with the enormously higher data flow we expected and realized. My experience with automatic radar tracking led me to propose that we follow and record our tracks automatically. A computer could then reconstruct them in three dimensions. It turned out that computers could also identify the mass and charge of each track. This great achievement, which I hadn't anticipated in my prospectus, was due to the efforts of Frank Solmitz and Art Rosenfeld and was quickly adopted by all bubble chamber users worldwide. Without these techniques the bubble chamber couldn't have made the significant contributions to particle physics that it did. (By 1968 we were measuring and analyzing some 1.5 million events a year.)

While we were building the 72-inch chamber, we were also conducting physics experiments with its smaller counterparts. Not often in physics has one group enjoyed a monopoly on an important frontier. Ernest Rutherford at the Cavendish enjoyed a monopoly on nuclear transmutations for several years after 1919. Enrico Fermi in Rome cornered the market on neutron-induced radioactivity for several months in 1934. We had the physics of negative K mesons (kaons) stopped in hydrogen all to ourselves for at least a year, which was more than enough time to mine the gold. The category sounds narrow; in fact it was one of the richest gold mines in the history of particle physics.

Our work with the 4-inch chamber at the 184-inch cyclotron and the Bevatron can't be dignified by the designation "experiment" but did show examples of pion-muon-electron decay and neutral strange-particle decay. Those demonstrations simply whetted our appetites for the exciting physics we felt sure the 10-inch chamber would manifest.

Bob Tripp joined the group in 1955 after taking his doctorate with Emilio Segrè. As his first contribution to our program, he designed a separated beam of low-energy kaons that would stop in the 10-inch chamber. We started our physics program on negative kaons stopped in hydrogen for two reasons. One involved physics: Pief Panofsky and his coworkers had shown that such interactions with negative pions were a rich source of fundamental knowledge. The other was pragmatic: only one straight section of the Bevatron was available for high-energy beam production by physicists, and it was consequently heavily oversubscribed, but we could make low-energy negative kaons from a target inside a curved section of the big accelerator. They bent away from the vacuum chamber and passed between the vertical iron members of the magnet yoke. So we had our own private source, which didn't interfere with other users and was available to us whenever the Bevatron was working. We were soon photographing stopped kaon reactions in pure hydrogen.

My scientific career had seldom been more rewarding. We not only had the world's finest detector; our stopped negative-kaon beam was also unique. We observed numerous reactions never seen before and measured masses, lifetimes, and decay branching ratios for a crowd of strange particles. We took full advantage of our outstanding bubble chamber photographs and wrote a number of definitive papers.

Our separated kaon beam—the first of its kind—was contaminated with pions and muons. That crudeness led to an interesting and seemingly portentous discovery. We wouldn't have made it if we had settled into full bubble chamber professionalism with clean beams and professional scanners carefully trained to recognize and record only certain criteria of events. We were still scanning our own film, so completely absorbed that someone was always waiting in line to take over one of our few film viewers whenever whoever was using it tired.

At first we kept records only of events involving strange particles. We looked quickly at each frame in turn (a roll of film carried four hundred stereo pairs) and moved on to the next if no such event occurred. But we began to notice what appeared to be an unusual decay scheme—a negative pion decaying to a negative muon decaying to an electron. The decay of a muon at rest into an electron was expected in hydrogen, but Panofsky and his associates had shown that a negative pion couldn't decay at rest in hydrogen. We thought at first that the pion had simply decayed just before stopping. Gradually we became convinced that this explanation didn't fit the facts. There were too many muon tracks of nearly the same length; if the pions had decayed in flight, the tracks should vary more in length than they did.

So we began keeping records of these anomalous decays, as we called them. We didn't yet have equipment for reconstructing tracks in three dimensions, but we sometimes found muons moving horizontally; we could measure the length of those tracks and thus estimate their energy. The negative muons, we found, had energies of 5.4 MeV, which was higher than it ought to be if their parent particles were in fact negative pions.

Then we made another surprising observation: we noticed a gap of a few millimeters between the end of a "pion" primary track and a muon secondary track. That implied that some intermediate neutral particle had traveled through the hydrogen; since it carried no electrical charge, it left no trail of bubbles behind. We had missed many pairs of tracks with these gaps because no one had seen such phenomena before; we therefore ignored them by unconsciously assuming that they were unassociated events in a badly cluttered bubble chamber.

One evening one of our team members, Harold Ticho, had dinner with a Berkeley astrophysicist—the two had been students together—and thoroughly went over our puzzle. The astrophysicist, J. A. Crawford, suggested that a nuclear fusion reaction might somehow be responsible. Fusion reactions, of course, power the hydrogen bomb. They are also the subject of continuing research in their controlled form as a possible source of energy for the generation of electrical power. They take place when two nuclei approach each other closely enough and stay there long enough to fuse together, with a corresponding reduction in mass and release of energy. Harold and Jack quickly showed that the fusion of a proton with a deuteron would supply just the energy we needed to explain our anomalous events.

A stopped negative muon, it seemed, might substitute for an electron and bind a proton and a deuteron into a very tiny molecule, the nuclei of which would remain almost touching long enough to fuse into helium 3. That would deliver the fusion energy to the muon by a process known as internal conversion.

But protons outnumbered deuterons in our hydrogen chamber by five thousand to one; we couldn't understand why we saw this unusual reaction so often. The next day, after all of us had accepted the idea that stopped muons were catalyzing the fusion of protons and deuterons, our whole group paid a visit to Edward Teller at his home. After we introduced our observations and described the proposed fusion reaction, Edward explained why the reaction had such a high probability. The stopped muon, he suggested, radiated its way into its lowest possible orbit around a proton, at which point the electron drifted away. The resulting muonic-hydrogen atom, small and electrically neutral, had many of the properties of a neutron and could diffuse freely through the hydrogen. When it came close to the deuteron in one of the rare hydrogen-deuterium molecules, conditions of energy would favor the muon transferring to the deuteron. This new, heavier "neutron" might then recoil some distance as a result of the exchange reaction, thus explaining the odd gap. The final stage in the sequence, the capture of a proton by the muonic-deuterium atom to make a molecule, would also be energetically favored; a proton and a deuteron would thus be confined closely enough by the heavy negative muon to fuse quickly into a helium-3 nucleus; the liberated fusion energy would then kick the muon away from the helium-3 nucleus by the process known as internal conversion.

For a few exhilarating hours we thought we had solved mankind's energy problems forever. Our first, hasty calculations indicated that in liquid hydrogen-deuterium a single negative muon should catalyze enough fusion reactions before it decayed to supply enough energy to operate the accelerator to make more muons, to extract the necessary hydrogen and deuterium from the sea, and to feed the power grid. While everyone else had been trying to control thermonuclear fusion by heating hydrogen plasmas to millions of degrees, we seemed to have stumbled upon a better than break-even reaction that operated at minus 250 degrees Celsius.

More realistic calculations showed us we were short of the mark by several orders of magnitude (from 1 to 10 is one order of magnitude; from 1 to 100, two; from 1 to 1,000, three). In physics that's a near miss. In recent years there has been a great resurgence of interest in fusion reactions catalyzed by muons. If one uses deuterium and tritium, as Steven Jones does, rather than deuterium and hydrogen, the average number of fusion reactions catalyzed by one muon is about 150, and each fusion yields 20 MeV instead of 5—very near to break-even. In all our work, we saw only two muons that catalyzed more than one reaction (in both cases, two). So Steve is getting very close to paydirt.

By the winter of 1958 the 15-inch chamber had completed its engineering test run as a prototype for the 72-inch and was operating for the first time as a physics instrument. By now we had an intermediate-energy negative kaon beam of much higher quality than our low-energy beam had been. We were hoping to detect a particle predicted in the theoretical work of Murray Gell-Mann, the xi-zero, that should materialize along with a positive kaon to conserve strangeness. We terminated that experiment with only a single xi-zero candidate, but it was in fact the first ever seen. This was a real tour de force; we discovered a new neutral particle that decayed into two other neutral particles, and none of the three left a track in our bubble chamber. One could always get a laugh at a meeting by showing a photograph of a blank cloud chamber and announcing, "This picture shows a new neutral particle decaying into two other neutral particles." Our picture wasn't blank, because we saw an interaction vertex where the xi-zero was formed and we could tell where the secondary particles had gone because they each decayed into charged particles.

We saw many new kinds of interactions in our 15-inch bubble chamber. One of these revealed a negative kaon interacting with a proton to produce a lambda, which decays into a positive proton and a negative pion. The rate of production of these events nicely matched our increasing ability to use our new computerized measuring machine, called the Franckenstein after its designer, and we soon had computer printouts that listed the energies of the four interesting tracks for each event.

Two of our graduate students then made scatter plots showing the energies of the two pions (a single dot on graph paper for each event, with its horizontal position determined by the energy of one of the pions and its vertical position determined by the energy of the other). To their surprise they discovered that the density of points didn't vary smoothly from one part of the diagram to another, as expected, but was crossed by two dark bands, one running horizontally, the other almost vertically.

That was an exciting observation. It indicated that what we had expected to be a three-body interaction—a lambda plus two pions—was in fact a sequential pair of two-body interactions: a pion recoiling against a heavy object that very quickly came apart into a lambda and a pion of the opposite sign. It was easy to calculate the mass of this heavy object to be 1385 MeV. We christened it the Y*(1385) (pronounced Y-star 1385). As we accumulated more data, we were able to use Heisenberg's uncertainty principle to measure its lifetime, which turned out to be about 10^{-22} seconds, making it a very short-lived object (at the speed of light, it can just cross an average-sized nucleus before it comes apart).

We soon discovered many similarly short-lived counterparts to the

Y*(1385). I was surprised that they should appear in a bubble chamber that was designed to look for particles that remained intact a million million times longer, the strange particles with lifetimes in the 10^{-10} second range. It was as if a fisherman with a net designed for catching sharks were catching minnows, except that the difference in particle lifetimes was of far greater magnitude, 10^{12}, than the difference in size between a minnow and a shark; the length of the minnow at an equivalent scale would have to be about 1 percent of the diameter of an atom.

No theorist had suggested that we look for such particles, which put our discoveries in a different category from those for which theorists supply detailed road maps. The search for the particles theorists predict is important, but it's like searching for the source of the Nile or pushing on through hardship to the South Pole; our findings corresponded to James Cook's coming in complete surprise upon the Hawaiian Islands, which no Westerner knew about or predicted. Important discoveries in geography and in physics turn up in many shapes and sizes; our short-lived resonance particles had special significance.

My name had not appeared on a single paper describing the development of bubble chambers. It properly appeared on the xi-zero paper, since I had proposed the reaction we used to make the particle and the technique we used to prove we'd seen it. But when we wrote up the Y*(1385), the first of three strange resonances we discovered using the 15-inch chamber, I removed my name from the head of the alphabetical list of authors and tucked it away in the final paragraph, in the thank-you note. I had received what seemed to me adequate attention for publications from the 10-inch chamber, with my name first alphabetically. I felt comfortable being listed first for my part in working out the K-meson physics and in being remembered for the development of the bubble chamber itself, but I had no exceptional claim to the discovery of the resonances. In removing my name from the list, I did what I thought Ernest would have done under similar circumstances. My young colleagues weren't able to persuade me to change my mind until they pointed out that Margaret Alston, a new member of the group, had been promoted from the thank-you list to lead author. My name would thus appear second, not first; the papers would be cited as "Alston et al."; so I agreed to be raised again to authorship.

My guardian angel was working overtime: without those three papers my name would have appeared on none of the work for which I was awarded the Nobel Prize in physics in 1968. If, as rumor has it, the Nobel committee can't award a prize to a person whose work doesn't carry his name, I might not have won the prize at all. A bookie who counted pages in the phone book would say that for Margaret to have a last name ahead of mine alphabetically would be a 42-to-1 longshot. I've frequently thought of the two graduate-student discoverers, Stan Wojcicki and Bill Graziano, as my Jocelyn Bells. Miss Bell was a graduate student in Tony Hewish's group and found the first pulsar, a discovery as unexpected as as the Y-stars. Hewish and I both received high honors for our continuing efforts to build the detection equipment that made these discoveries possible.

The 72-inch bubble chamber operated for the first time on March 24, 1959, nearly four years after it was first seriously proposed. It served well; in the years to come, following Ernest Lawrence's tradition, we supplied our physics colleagues all over the world with some ten million high-quality triad stereo photographs. "Practically all the discoveries that have been made in this important field of high-energy physics," a member of the Swedish Academy of Sciences noted at my Nobel presentation, "have been possible only through the use of methods originated by Professor Alvarez." I cite this comment not from immodesty but to emphasize how far into physics our Berkeley bubble chamber work reached.

The day before the manuscript for this book was to be delivered to the publisher, I read a review in *Nature* of Rudy Peierls's autobiography, *Bird of Passage.* The reviewer concluded, "I should have preferred to have seen somewhat more of Sir Rudolf's views on the state of physics and less of personal anecdote." To protect my book from roasting by reviewers with similar tastes, and because they might even be important, I offer the following observations.

There is no way that a person with my personal qualities could go into either nuclear physics or particle physics at the present time. The table of contents of the latest issue of *Physical Review Letters* lists three particle physics papers with multiple authorships. The first two papers each have 72 coauthors, taking up fifteen lines—exactly the same names, in the same order—and the third paper lists 46 coauthors. I can't believe that I could ever have derived any satisfaction from being listed as the 37th in a group of 75, or as the 337th name on the list of 500 that will soon characterize the papers coming from the large European electron-positron colliding-beam accelerator near

Geneva. I once saw a cartoon over the desk of a person working in one of these huge collaborations; it showed two men chained to a trireme oar, pulling as hard as they could. One said to the other, "If it weren't for the honor of the thing, I'd rather be doing something else."

My observations of the young physicists who seem to be most like me and the friends I describe in this book tell me that they feel as we would if we had been chained to those same oars. Our young counterparts aren't going into nuclear or particle physics (they tell me it's too unattractive); they are going into condensed-matter physics, low-temperature physics, or astrophysics, where important work can still be done in teams smaller than ten and where everyone can feel that he has made an important contribution to the success of the experiment that every other member of the collaboration is aware of. Most of us do physics because it's fun and because we gain a certain respect in the eyes of those who know what we've done. Both of those rewards seem to me to be missing in the huge collaborations that now infest the world of particle physics.

My second criticism of the way particle physics is done these days is that the theorists have veto power over almost any experiment proposed by their experimental colleagues. The last completely unexpected discoveries were made twelve years ago, at Stanford and at Brookhaven, of the J/ψ meson and the tau lepton. I described my debt to Hans Bethe in earlier chapters, so I can't be accused of any anti-theorist bias. I think theory and experiment have been equally important in the evolution of our science. My complaint is only that at present almost all particle physicists feel that an experimentalist's only function is to check the predictions of theorists. Witness the last great discovery in the field—the existence of the W and Z particles—at CERN, Geneva. Every theorist was convinced, before the discoveries were made, that the W and Z particles existed. Carlo Rubbia and his many collaborators confirmed the approximate masses. What was really learned?

An answer to that question may be found in the many recent, very expensive, and completely unrewarding searches for proton decay, also predicted by theorists. No one would have dreamed of setting up an experiment just to look for proton decay, but as soon as some of the best theorists said the proton would almost certainly decay, with a measurable half-life, the funding agencies were flooded with proposals to search for the decay. By now, we know the half-life is more than one hundred times longer than the predicted value and quite consistent with infinity, the "old value."

I must reiterate my feeling that experimentalists always welcome the *suggestions* of the theorists. But the present situation is ridiculous. Theorists now sit on the scheduling committees at the large accelerators and can exercise veto power over proposals by the best experimentalists. Mel Schwartz was so outraged by this situation that he gave up his tenured professorship at Stanford and has gone into business for himself, where he feels he again has some control over his own destiny.

In the days when I was active in nuclear and particle physics, the theorists exercised their veto power in an acceptable manner—they held up to ridicule anyone who did "stupid experiments." So if anyone could be intimidated by such social pressure, and nearly everyone can be, they were effectively kept from doing stupid experiments. Occasionally an experimentalist did such an experiment without realizing that it was stupid and published a stupid result, to find himself ridiculed by the much smarter theorists. A classic example was the experiment in which R. T. Cox found, in 1928, that the electrons emitted in the beta decay of radium E must be polarized—because they "double-scattered" with different probabilities to the left and to the right. Theorists couldn't accept this result, because it violated the principle of the conservation of parity, which they held to be sacred. So the important Cox discovery disappeared from sight until it was rediscovered "properly." Nineteen years later, T. D. Lee and Frank Yang, to solve a serious puzzle in particle physics, proposed that parity might not be conserved in weak interactions. Now that theoretical holy water had been sprinkled on a search for parity violations in weak interactions, several teams set up experiments to look for it, and Lee and Yang's suggestion was quickly found to conform to the structure of the real world. But the Nobel Prize for this change in paradigm went neither to Cox nor to Madame Wu and her collaborators, who proved Lee and Yang to be correct. Lee and Yang certainly earned their prize, but I believe the experimentalists should have shared in the glory.

A few years before the law of parity conservation was properly overthrown, Norman Ramsey told me about what he characterized as a close brush with disgrace. It occurred to him that he could use the 100-percent polarized beams of neutrons that Fermi was producing by reflection from magnetized mirrors. He would let those neutrons be captured by some nucleus with zero spin, such as neon 22. That would produce, according to Norm Ramsey, polarized neon-23 nuclei, as he could verify by seeing the up-down asymmetry in their emission of beta rays. He told me how he had called Enrico to see if he could use the fancy neutron beam. "You know," he said, "I've never been so lucky in my life; Enrico was teaching a class, so I couldn't talk with him, and by the time the class was over I had realized that the up-down asymmetry of such polarized nuclei was forbidden by parity conservation. I wouldn't have enjoyed being told by Enrico that I was proposing a nutty experiment, even though he would have let me down gently." This conversation is reconstructed after about thirty years, and some of its physics may be incorrectly remembered. But the relief Norman experienced in escaping a painful lesson in theoretical physics from his friend Enrico Fermi was palpable and never to be forgotten. (That's what I mean by control by ridicule.)

Next year is the one-hundredth anniversary of the Michelson-Morley experiment. That terribly important experiment couldn't be done at the present time, as I think the following imaginary scene will show. Michelson and Morley tell the "scheduling committee" that they plan to measure the velocity of the earth through the ether by means of Michelson's new interferometer. The theoretical astrophysicist on the committee asks, "How accurately will your method measure the velocity?" Michelson says, "To about one significant figure." The theorist responds, "But we already know the velocity of the earth through the ether from astronomical observations, to two or three significant figures." All members on the committee agree that the proposal is one of the nuttiest they've ever heard, and Michelson and Morley find themselves turned down flat. So the experiment that led to the abandonment of the concept of the ether isn't done.

I wonder how many physical concepts that "everyone knows" to be true, as everyone knew parity conservation and the existence of the ether were true, would turn out to be false if experimentalists were again allowed to do nutty experiments. This impasse is not a problem only in physics. What would the peer reviewers in the National Science Foundation's paleontological section have said about a proposal from a Berkeley group to look for evidence that an asteroid or comet impact led to the extinction of the dinosaurs sixty-five million years ago?

In my considered opinion the peer review system, in which proposals rather than proposers are reviewed, is the greatest disaster to be visited upon the scientific community in this century. No group of peers would have approved my building the 72-inch bubble chamber. Even Ernest Lawrence told me he thought I was making a big mistake. He supported me because he knew my track record was good. I believe that U.S. science could recover from the stultifying effects of decades of misguided peer reviewing if we returned to the tried-and-true method of evaluating experimenters rather than experimental proposals. Many people will say that my ideas are elitist, and I certainly agree. The alternative is the egalitarianism that we now practice and that I've seen nearly kill basic science in the USSR and in the People's Republic of China.

I don't expect to see much change in our disastrous peer review system, since people always ask, when they hear of my proposal, "How would a young scientist ever get any financial support? He wouldn't have a track record to be evaluated." I think he would do it the same way I did; I apprenticed myself to Ernest Lawrence, a man with such a good track record that any proposal he made was automatically funded without peer review. After a few years, if he had any talent, my young scientist would have his own track record and assurance of support when his peers reviewed *him*—not his proposal.

Perhaps the most serious defect in the present peer review system is that it effectively prevents scientists from changing fields. So I see lots of my friends spending their whole scientific lives essentially repeating their Ph.D. theses. If they find that activity too dull, and want to move into a different field, their longtime funding agent tells them, "I'm sorry to see you leave this field, and of course I won't be able to support you in your new activities—they aren't in my department." That's enough to keep almost anyone from leaving his own sandbox, where he was so well protected for years, and moving into a neighboring one, where the occupants will see to it that he doesn't get any of their hard-won toys.

TWELVE

The Prize

IN APRIL 1956 the Soviet Academy of Sciences invited fourteen U.S. physicists to attend a high-energy-physics conference in Moscow. Stalin had died only three years earlier; Nikita Khrushchev's speech denouncing him at the Twentieth All-Union Party Congress was fresh news. If someone had suggested before the invitation that I would soon be in Moscow, I would have replied that a visit to the back side of the moon was more likely. We were to open the USSR to Western scientists.

I kept a 20,000-word diary during that trip, intending it for family and close friends; it was subsequently syndicated in an edited version by the United Press and published in *Physics Today.* I find those exciting and unexpected events less interesting now than they seemed at the time. Soviets, British, and Americans alike, all of us were happy to be together and to talk of our work during and after the war, and we were careful not to say anything that might chill the burgeoning fellowship between our countries. Khrushchev's speech, for example, had been published in the West but not in *Pravda;* we didn't mention it to our Soviet colleagues.

I had a suite overlooking Red Square and the Kremlin, with a piano; I was undoubtedly the first to play "Davy Crockett" on that piano. For theorists the high point of the visit was the chance to meet Lev Landau, the outstanding theoretical physicist who would win the 1962 physics Nobel Prize and who had been jailed in 1938, in one of Stalin's purges, until Peter Kapitza succeeded in forcing his release. For experimentalists, certainly for me, the high point was a chance to tour the previously secret Dubna laboratory, several hours north of Moscow, where the Soviets had built a beautiful 680-MeV cyclotron. I noted in my diary,

> The briefing . . . was impressive, but we were bowled over when we saw the cyclotron itself and all the experimental apparatus which was set up for business. . . . I have seen all the large American cyclotrons and this is better engineered than any of ours. But even more impressive than the accelerator itself was the amount of experimental equipment, its technical excellence from the engineering and physics standpoints and the number of experiments set up simultaneously. They have sixteen different beams.

At the same time, I noticed then what has proved to be true of many aspects of Soviet science:

> Although I heard of some very novel and even brilliant ideas in the field of machine design, I didn't see or hear of a single new idea for an experiment in nuclear physics. I feel sure that some of the new crop of experimental physicists will be able to think of ideas for new experiments, but I don't know how favorable the climate will be for them to try them out. . . . The beautifully set up series of experiments at the 680-MeV machine is most impressive to a Westerner, but after some thought it might indicate a bit too much regimentation. The touch of anarchy we see in our labs is really a good thing for the advance of physics. No director can think of everything, so he must trust his young people to try their own ideas, even if most of them are no good.

We toured the Kremlin; I was surprised to be taken into Lenin's private apartment and office:

> This room is one of the sacred shrines of Communism and our interpreters were absolutely bug-eyed to find themselves there. Lenin worked at a small desk, sitting in a straight chair. The desk butted up against a felt-covered table surrounded by large stuffed-leather easy chairs. The table and chairs were for the visitors and government officials who came to talk with him. The room was roughly square and about twenty or twenty-five feet on a side. The wall behind Lenin's chair was covered with bookshelves which were filled with books in Russian, French, German and

> English. Lenin read all these languages and was apparently a rather scholarly man. . . . We walked around the room and examined the various objects on the tables. I had the feeling that Lenin might show up at any minute—the room was in such a natural state. There was a large oil painting of Karl Marx on the wall, facing Lenin as he sat at his desk. The maps had been marked by Lenin, and the calendars on the desk and on the wall were opened to a day in 1923.

Kapitza gave a party for us at his institute after we'd all had time to become acquainted:

> Several long tables were set up in his large office, and we had a nice supper with lots of wine, caviar, sandwiches, cheese, and fruit. Before the supper, we got into a typical physicists' session of puzzles and jokes, so we were relaxed. This party was the most natural one of the conference—both Russians and visitors have gotten quite used to being with each other, so it no longer is the case that everyone is consciously trying to be on his best behavior. We just acted naturally and had a wonderful time. There was lots of laughter and some hot arguments about physics between the theoreticians—just what goes on at any gathering of American physicists. Everyone knew that we were happy to be here, and that the Russians were happy to have us as their guests.

The laboratories had old-fashioned hardwood floors, and even the new buildings sported Victorian chandeliers, odd incongruities among cyclotrons and motor-generator sets. The wife of the director of the Moscow Physical Institute had read *Gone with the Wind* twice, I learned from her at a dinner, and lamented the absence of good Russian detective novels. At the Hermitage, that extraordinary museum, I found the four walls of one room covered with Tahitian-period Gauguins. Russian nobles had bought them in Paris, and the Bolsheviks had confiscated them.

Despite the elaborate banquets and extensive tours we were ready to leave:

> We had a good flight from Riga to Stockholm. It was wonderful to see a bit of the Western world again, in the shape of an airplane that was spick and span inside, with good service and well-dressed people. The plane was full of late copies of *Time, Life, Look,* and English and American papers. We devoured the literature like starving men. As we flew over Stockholm it was quite apparent that we were in a different civilization. Everything looked clean and neat, and cars were plentiful on the roads. And then when we drove through the streets—the women! Sweden is famous for its beautiful and well-dressed women, but I am sure they never

have such a appreciative audience as they find in a plane load of Westerners from the Soviet Union.

I had a well-earned rest in Stockholm, then flew to Geneva for the first truly international postwar conference on particle physics. On the opening day I found myself seated at lunch next to Rolf Wideröe, the inventor of the drift tube RF linear accelerator whose diagrams had catalyzed Ernest Lawrence's ideas for the cyclotron. Since I had invented a new kind of linear accelerator that accelerated protons, Wideröe and I had much to talk about. Ernest was sitting nearby. During our conversation I asked Wideröe if he had ever met him. He said he hadn't, and I had the pleasure of introducing them.

Most of the talks at Geneva I had already heard in Moscow, so I flew back to San Francisco after only a day there. When I arrived home, it was clear to both Geraldine and me that our marriage was failing. We had an unspoken agreement to remain together until the children were grown; we almost made it. After twenty-one years, when Walt was a senior in high school, Gerry and I were divorced.

My closest friend then was Rowan Gaither, in the spring of 1957 president of the Ford Foundation, Ernest's Chromatic Television partner, and my Bohemian Grove camp mate. Once a year Henry Ford II sent Rowan a sheet of paper with ten numbered lines, on each of which he was encouraged to write the name of a friend who could then purchase any Ford automobile at factory cost. Because Rowan always included my name, I became known for driving a new Lincoln convertible, a practice that continued even when I was very pressed financially, paying more than half my salary for alimony and child support. The car was the cheapest Ford product I could buy; I often sold last year's model for more than the new one cost. (After Rowan died, I bought a secondhand Falcon, equally the cheapest Ford product then available to me.)

In early June, Rowan was scheduled to speak at the annual dinner of the American Society of Mechanical Engineers in San Francisco. He invited me to attend. Jim Landis, the president of the ASME, had met Rowan on a plane; I found myself that evening seated next to Jim's daughter Janet, who had just been hired at the Berkeley Radiation Laboratory and had many questions to ask. Jan was charming and attractive, and we talked of the things that interested us. My only thoughts of romance that night extended to guessing which of the young men at the laboratory would snatch her up. Some of my friends' wives had sternly warned me not to involve myself with young women. Jan was then twenty-six years old; I would be forty-six the next day.

Jan started at the laboratory scanning nuclear emulsions in Walter Barkas's group. That was dull work; I persuaded Walt to let her scan bubble chamber film. She fit in well and soon became a computer operator and then supervisor of our large scanning group.

In the meantime I departed for Washington for two months' work on the Baker panel looking into the activities of the supersecret National Security Agency. I was worried that I might not be clearable for such a sensitive project, because over my desk I had a photograph of myself standing at the podium of the Supreme Soviet, deep inside the Kremlin, with a statue of Lenin behind me and the hammer-and-sickle plaque attached to the dais. The security man assured me that he knew of the photograph and that I had already been cleared. We were the first outside panel with access to NSA secrets; the exceptionally competent, mathematically oriented NSA scientists were eager to tell us about their achievements. I especially enjoyed learning in great detail from William Friedman how the United States broke the Japanese diplomatic codes before World War II.

I spent the Independence Day weekend with Alfred Loomis and his second wife, Manette, in East Hampton on Long Island; when I returned to Berkeley in early August, I saw Jan daily at the laboratory and at an occasional party. I still imagined no romance. But after dinner on August 16, while I was working in my office and Jan was studying computer printouts in the next room, our friend William Nierenberg dropped by and asked if we had seen the comet. We hadn't. Bill insisted we accompany him to the roof. We did; comet Mrkos, with its tail extending far above Mount Tamalpais, was a lovely sight. We watched it for some time. Eventually Bill excused himself. If the subsequent events had been part of a movie, the camera would have panned from Jan and me, silhouetted against the lights of the city, to a sky filled with star bursts and fire falls. We still celebrate Comet Day and ASME Day, private holidays, every summer.

We were discreet about our newly discovered love; for several months we imagined we had fooled everyone. We didn't fool Bill, who arrived home that night from comet viewing and told Edie, his wife, to start looking for our wedding present. Jan and I were married in her parents' home on December 28, 1958, with close friends and relatives attending and with Don Gow as my best man, as I had been his.

Jan and I spent our honeymoon at the Eldorado Country Club in Palm Springs, to which President Eisenhower retired when he left Washington. Golf was my main recreation for many years, and one of my golfing companions was Ed Janss, who, with his brother Bill, bought Sun Valley from the Union Pacific Company after previously founding Snowmass at Aspen. I had been his partner on a number of occasions at the members and guests tournaments that make life in the desert so pleasant. He offered us his beautifully appointed "cottage" at the tenth tee at Eldorado next to the fabulous clubhouse that is known in the golfing world as the Taj Mahal. It would be hard to imagine a more unlikely setting for an impoverished professor and his bride to start their life together.

Two "Rochester" conferences in 1959 and 1960 gave me further exposure to the people and the politics of the USSR.

The 1959 conference was held in Kiev. We knew a year in advance that we would go, which gave Jan time to take a crash course in spoken Russian. During the many trips we took in the first months of our marriage, she worked with packs of flash cards to improve her vocabulary. It stood us in good stead. Almost all the Soviet physicists spoke excellent English, but their wives often did not. Nor did hotel and store clerks.

Serge Nikitin and his wife met us at the Moscow airport. In many ways Serge was my Soviet counterpart. He built the first Russian hydrogen bubble chamber and at the time of the conference was planning a larger chamber. After my 1956 visit to the USSR he had visited Berkeley; in long talks together we had become close friends.

The Nikitins saw us through customs—our bags weren't opened—drove us to the new Ukraine Hotel, and turned up the next morning to show us the sights, many of which weren't on the Intourist route. Jan had brought with her to Moscow a small tape recorder with which to dictate a daily account of our experiences. After dinner at our hotel, back in our room, she found the tape recorder inoperable. I correctly surmised that our bags had been searched and the recorder disabled while we were out.

After several days of sightseeing in Moscow we boarded the special Soviet Academy sleeping car attached to the train for Kiev. I knew all the male passengers except the Russian who shared our four-bunk compartment. He responded neither to my English nor to Jan's Russian, so I assumed he was KGB. We were unhappy to share a sleeping compartment with a stranger; when dinnertime approached I pointed out several unoccupied compartments to Serge and asked him to help us get rid of our unwanted guest. "Don't worry," Serge said; "I'll talk to him and he'll be glad to move." When we returned to our compartment after dinner the man was still there, and he remained there overnight. For anyone who has outgrown the tooth fairy, that should confirm my guess at his identity.

I probably saw less of Kiev than anyone who attended the conference. The meeting format had been changed at the last moment; the contributed papers were to be consolidated and put into perspective by an expert. The organizers insisted on a finished manuscript before the conference opened so that they could publish the proceedings promptly. As the first particle physics rapporteur, I spent most of the week stuck in my Kiev hotel room. I worked day and night, leaving the room only for meals. No one was allowed to insult the waitresses by tipping; the service in the dining room was correspondingly slow, and Jan and I confined ourselves to two meals a day in order to get anything else done. Once a day a sad-looking English-speaking secretary would knock on our door, collect the latest pages of my handwritten manuscript, and deliver a typed copy of the preceding day's work. Since she could transcribe my scrawl, I concluded that she had been raised in an English-speaking country. I tried to imagine how the poor woman had been trapped into living in Kiev, but I couldn't invent a reasonable scenario and didn't have the heart to ask her.

Consolidating the various papers I was responsible for put a severe strain on my friendships. Everyone remembered something I had failed to mention. I had, in particular, a vast amount of data from nuclear emulsions. I had dismissed that outmoded recording technique publicly as equivalent to strawberry jam because particle interactions took place in emulsions not with simple hydrogen nuclei but with complicated objects like carbon, oxygen, silver, and bromine nuclei. Among the souvenirs someone presented me during this period was a jar of strawberry jam with an Ilford Nuclear Emulsion label on it. I told the emulsion people that I didn't feel competent to sort out what they had sent me and would stop my presentation ten minutes early to let one of them report, a compromise that didn't please anyone.

Once the conference began, I was liberated. Jan and I enjoyed strolling across the center of town each day to the meeting hall. We frequently walked with George Volkoff, who had been one of Robert Oppenheimer's students at Berkeley before the war. George was Canadian, but his father, an Orthodox priest, had insisted that the family speak Russian at home. From George we learned what the citizens of Kiev were saying to each other. He couldn't get used to so many people speaking his secret family language. It had been spoken politely in his home; he was shocked to hear it used rudely.

Rudeness was the order of the day; my sessions chairman, a Chinese doing research at Dubna and the first Chinese Communist I had ever met, must have been instructed to be rude. (All the Chinese Jan and I have met on our two trips to the People's Republic have been very friendly.) I told him that one of my colleagues would use the last ten minutes of my allotted hour. Thirty-five minutes into my talk he interrupted to inform me that my time was up. I said I had another fifteen minutes and continued. Five minutes later he repeated his challenge. I corrected him again and continued. Five minutes after that he did the same thing again. This time I threw up my hands and stepped aside. The man took an unscheduled twenty minutes to announce the discovery of a new particle in the Dubna propane bubble chamber. I had never seen such an amateurish performance before. The new particle was never mentioned again.

When people asked me later how the Russians were doing in particle physics, I usually answered that they had discovered minus five particles. The brothers Alikhanov claimed three different kinds of "varitrons" in cosmic rays and a heavy relative of the muon; my Chinese chairman had his Dubna discovery. In a country with a well-behaved scientific tradition, the claimers of such particles would eventually retract their claims, which no one else could verify. But the Alikhanov brothers, for example, received the Lenin Prize for their unverifiable discoveries; for them to have retracted their claims would have been embarrassing to their government.

I knew, liked, and admired many talented Russian particle physicists, and I pitied them because of their working conditions. They worked in a centrally planned system that was ill designed to cope with the uncertainties that set the research world apart. Serge Nikitin's head start in liquid-hydrogen bubble chambers, not to mention his enthusiasm and talent, should have made him a good competitor of ours. But his large chamber was never completed, or at least never produced any results. I didn't ask Serge why, because I didn't want to embarrass him. I was told many years later that someone at Dubna had decided that liquid hydrogen was too dangerous and had forbidden such chambers at their accelerator.

To the best of my knowledge, no Russian high-energy accelerator has ever accomplished an important experiment. The 10-GeV Synchrophasatron at Dubna was a failure; its designer, a first-class physicist, didn't realize that iron supplied from several different mills had to be shuffled before being assembled into magnets to average out its differences. The masonry wall that surrounded the machine confined it so closely that it left no room for detection devices. The Soviet government seemed to think that accelerators were important in themselves; it apparently didn't understand that such machines were only tools for doing physics.

The highlight of the Soviet soccer season that year was a match between the Moscow and Kiev teams, but our Intourist guides scotched our requests for tickets with their most overused word, "*Eem*-possible." On our last day in Kiev we were surprised to find that our guide was ready to deliver us to the match. The stadium was already crowded; we had no tickets; but the woman herded us to the center line and created seats by motioning to people to make room for us. They didn't protest, and we watched a fine soccer game, after which our guide escorted us back to our hotel.

It was the cocktail hour; I headed for my large suitcase to set up the bar. The suitcase had a broken clasp that I always caught with my left hand as I released it with my right. This time the clasp didn't fall away. Someone had fixed it. I guessed that the KGB wanted to confirm that I hadn't repaired our tape recorder. I hadn't, but their concern earned us two ticketless seats to a great soccer game and a repaired suitcase clasp. They must have assumed they broke it. (We met Gilbert and Helen Highet at the Loomises' on our return to the United States; she writes spy thrillers under her maiden name, Helen MacInnes. She asked if she could use the clasp incident in her *Decision at Delphi* and did.)

We flew back to Moscow on a Tu–104, the world's first passenger jet; it was a converted Badger bomber, just as our Boeing 707 was a converted KC–135 military tanker. We took off from a military airfield parked with Badgers. The stewardess patrolled the aisles and ordered everyone not to photograph the planes. It hadn't occurred to me. My physicist's instincts aroused, I got out my camera. Whenever the stewardess retreated to the rear compartment I snapped a picture. At Strategic Air Command headquarters in Omaha a few months later (as a trustee of the MITRE corporation), I took along an enlargement that displayed four Badgers wing tip to wing tip. I showed it to a SAC officer. "Where in the *hell* did you get this?" he asked.

The next year's Rochester conference was held, for a change, in Rochester, New York. Jan and I spent a good deal of time there with Serge Nikitin and a colleague from his institute. During a weekend break we decided to show our friends something of the United States. We rented a car, turned over the road maps, and asked them to navigate: we would see what they wanted to see and go where they wanted to go—Niagara Falls on Saturday, they chose after due consideration, Cornell on Sunday.

They were excited by the adventure. Halfway to the falls they asked us to leave the freeway and drive through a small village on a lake. After touring its main street, we parked and explored. At the falls we rented a plane and toured by air, Jan surprising our friends by doing most of the flying. We took Serge and his colleague to a bowling alley that night; Serge was elated to make a strike. At Cornell the next day we bought Frisbees, instruments unfamiliar to our visitors, and sailed them on a deserted Cornell athletic field.

As we drove back to Rochester through an industrial town—a factory on one side of the road, houses on the other—Serge asked me how much these workers' houses would cost. I said I had no idea but would scout a For Sale sign and find out. We located a house for sale a few blocks farther along, parked, and trooped together to the front door. The yard was run-down; I was afraid our visitors would take home a poor impression of how our downtrodden workers lived.

The lady of the house was happy to give us a tour. She quoted a modest purchase price. The first-floor living room had wall-to-wall carpeting, comfortable chairs, a sofa, and a large color television set. The modern kitchen contained a big refrigerator, a freezer, an electric stove, and a dishwasher, all well maintained. After we saw the basement with its washer and drier, the woman asked us if we would like to see the upstairs, where her daughter and son-in-law had remodeled some of the rooms into an even more handsome apartment. After our tour we thanked the owner and drove on.

Serge and his colleague were stunned. They had just inspected a worker's home, apparently randomly selected, that was more elaborately furnished than many in which highly placed Soviet professionals lived. For the next several miles our guests discussed what they had seen in animated Russian. I'm sure they were asking what one always asks after seeing a magician do an apparently impossible trick, "*How* did he do it?"

I strongly dislike the Soviet system, especially for the frustration it visits upon the lives of its fine, talented scientists. But I deeply admire people like Serge Nikitin. I doubt if I could find the inner strength to tolerate their system. On the other hand, none of them ever gave me any indication that he disagreed with it.

Just before our Russian guests had to leave for home, they were entertained in Berkeley by Carl and Betty Helmholz. Betty comes from a wealthy family, so their home is large and elegantly furnished, with a swimming pool and a guest house. One of the guests couldn't believe what he was seeing and asked me if all American physicists lived this well. I replied with the truth, nothing but the truth, but not the whole truth, and in the process completely satisfied my questioner. "No," I said, "but Professor Helmholz is *head* of the physics department." (Heads of Russian institutes live better than any of their subordinates.)

Several traumatic events marred these otherwise happy years.

Ernest Lawrence died at the Stanford Medical Center on August 27, 1958, without recovering consciousness after major surgery. He was only fifty-seven; his untimely death shocked his wide circle of friends in science, government, and industry. He had led a life of enormous usefulness to science and to his country, his influence effected largely by example and by strength of character. He greatly admired scientific colleagues who could influence national policies through high administrative office or by public writing or speaking. But he felt that his own proper way was to advise from behind the scenes. Leaders in government and industry respected his distaste for the limelight and sought his counsel, knowing he represented no one but himself. His place in the history of science is secure. He will always be remembered as the inventor of the cyclotron. More significantly he should be remembered as the inventor of the modern way of doing science. When his young associates discovered element 103, shortly after his death, they named it lawrencium in his honor.

In 1960 Geraldine called from Chicago to say that our daughter Jean, then fifteen, had been hospitalized with severe chest pains. X-rays revealed a large, rapidly growing lung cancer. Jan and I caught the next plane to Chicago and spent the evening with Jean in her hospital room; she was scheduled for surgery the following morning. We waited then with her mother and her uncle to hear the surgeon's report. It was grim: he found cancerous growth so completely engulfing Jean's heart that he was unable to remove it.

The only remaining treatment was radiation therapy. My radiologist friends judged the case nearly hopeless. Their qualification concerned the cancer's rapid growth, which should, they thought, make it more vulnerable to radiation than the heart and surrounding tissue. They were right; massive doses of gamma radiation from cobalt 60 created in a nuclear reactor shrank Jean's tumor until it disappeared. Twenty-five years later she continues alive and well, I'm happy to say. Her first work after graduating from college was instructing physical education at Wellesley College. Later she earned a Ph.D. in psychology and became an active consultant.

Following Ernest Lawrence's death the president of the University of California, Clark Kerr, whose wife, Kay, was my stepcousin, said he would like me to be the new Radiation Laboratory director. I said I was sorry, but I was too involved in my work with the hydrogen bubble chamber to be effective. Glenn Seaborg, recently appointed as Berkeley chancellor, would have been ideal. Though Ed McMillan had avoided management roles throughout his career, he was the only other distinguished scientist available. I encouraged Clark to appoint him and said I would help him in any way I could.

My relationship with Ed, which had been very solid for years, both in and out of the lab, began deteriorating within a few months of his appointment. Eventually it led to our professional estrangement, a problem I solved by resigning as associate director of the laboratory and as head of the Alvarez group, since known as Group A, and forming a new group that received independent funding from NASA.

A more tragic outcome of the conflicts at the Radiation Laboratory was Don Gow's resignation. He left the laboratory where he had spent his entire adult life and took over the presidency of a nuclear-instrumentation company. For a while the company did well, but eventually it fell on hard times. Depressed, Don killed himself.

The last traumatic event of that difficult year occurred one day in October when the 1960 Nobel Prize was announced. Glenn Seaborg, who shared with Ed McMillan the 1951 prize in chemistry, had recommended that Don Glaser and I share the physics prize. Don, who moved to Berkeley in 1959, had invented the bubble chamber, he argued, and I had developed the hydrogen chamber and shown that it could be used to do important physics. There was a precedent: C. T. R. Wilson had won for inventing the cloud chamber and Patrick Blackett for developing it and using it, though their prizes didn't come in the same year. I thought that someday Don and I would share the platform in Stockholm, but I wasn't holding my breath.

Late one afternoon Glenn called me and asked me what I thought of the news. "What news?" I asked. That Don Glaser had won the Nobel Prize, he said. I told him I hadn't heard but that I would find Don immediately and congratulate him, which I did. It was soon obvious from the number of champagne bottles already emptied that I was about the only person in the lab who hadn't heard the news. None of my friends had felt up to the task of informing me that I had been considered by the Nobel committee and found wanting. That is something few physicists ever hear; nor have many seen that judgment reversed.

I've enjoyed my life immensely, but I would not like to relive 1960.

On October 29, 1968, Jan and I attended a party given by the 1961 chemistry laureate, Melvin Calvin. The science editor of the *San Francisco Chronicle* introduced me there to his wife. "Oh, you're one of the Berkeley Nobel laureates," she said. I replied with my standard clarification that lots of my friends had won the prize but that I hadn't. That was the last time I had to apologize for not having won a Nobel Prize.

At 3:30 the next morning the phone rang. When I answered it, someone at CBS News in New York announced that word of my prize had just come over the wire from Stockholm. I asked with whom I was to share it. It was all mine, he answered. My immediate reaction was pleasure mixed with surprise. I had written off the possibility because I knew very well that those who are passed over seldom get a second chance. Jan was awake by then; we congratulated each other as the phone rang again. This time United Press was calling with the same wonderful news.

I hung up the phone. Jan and I talked over our reactions. We had both seen the prize change the lives of some of our friends. We liked our life the way it was and agreed there in the early-morning darkness not to change it. I made two resolutions. I wouldn't sign petitions, and I wouldn't accept social invitations that were tendered only because of my Nobel Prize.

Surprised that the phone wasn't ringing, I checked the handset and found it misplaced. As soon as I replaced it it rang. Jean was calling from Wellesley, where her alarm radio had awakened her with the news. From then on the phone rang constantly with congratulatory calls from friends. By 4:30 Jan and I were dressed; by 5:30 the photographers had arrived. Our first visitor, at about seven, was our neighbor Frances Townes, Charles's wife (physics, 1964), who told us what was in store for us in Stockholm. Soon the official telegram arrived. I was struck by its length, the longest citation in the history of the prize.

When I finally freed myself from the telephone and drove to the lab, I found my office hung with congratulatory signs among which floated helium-filled balloons labeled with the names of the ten or twelve new particles our group had discovered. Lunch with my colleagues in the cafeteria was the first relaxed moment of the day; upon our arrival all in the room stood, applauded, cheered, and beat their plates with spoons. Many of them had contributed, and they were partly and properly also congratulating themselves.

A few hours later I was sitting on the platform of the laboratory auditorium at a hurriedly arranged press conference chaired by my close friends President Charles Hitch and Chancellor Roger Heyns. The room was packed to the aisles; I had the pleasure of recalling for the crowd what my friends had contributed to making that day possible. In the course of the day, in my office, friends and well-wishers consumed forty-seven bottles of champagne.

The next day Ed McMillan, who had been away in Pasadena, gave a laboratory-wide party in the cafeteria. That evening I practiced my resolve not to let the prize influence my personal life by taking my son Don, then three years old, trick-or-treating, his first Halloween rounds, keenly anticipated. I drove him to the homes of friends and sat in the car with his one-year-old sister, Helen, while he trundled up to receive his treats. (When our children were in college—at MIT and Haverford—we attended the 1986 Nobel festival, where Don and Helen saw what they missed eighteen years earlier.)

Slowly things returned to normal. I received some 750 letters and telegrams. I had sometimes been disappointed to receive a printed thank-you card when I had written a friend to congratulate him on his prize; I answered all my communications by hand, though I didn't finish the job until the following April. Ed McMillan lent me his white vest, worn by most of the Berkeley laureates over the years at the Stockholm presentation ceremony. I was pleased to carry on the tradition. In November one of my colleagues asked me if he and his wife could arrange to attend the ceremony. I thought then about the other group members whose great efforts had made my prize possible. I would be awarded more than enough tax-free funds to buy all the senior group members and their wives round-trip tickets to Stockholm. I tried the idea out on Jan. She liked it. Eight couples finally found they could go. My son Walt and his wife, Millie, flew up from Libya, where he was working as an oil geologist, and altogether our party consisted of twenty people.

We were accommodated in style in the Grand Hotel with a view of the palace across a narrow finger of water. We would have dinner there the day after the award ceremonies. The manager reminded us to be up and dressed on Lucia Day, when the most beautiful girls in Stockholm would serenade us wearing white dresses and crowns of lighted candles, an old Swedish custom. Harvard's Percy Bridgman, the 1946 laureate, had forgotten that appointment and was lying naked on the floor doing bicycling exercises when the manager let the Lucia girls into his room.

The award ceremony began at 4:30 in the afternoon, all the seats in the Grand Auditorium of the Stockholm Concert Hall filled with men in white tie and tails and women in formal evening gowns. Relatives of the candidates were seated in the front row just ahead of the diplomatic corps. Because Alfred Nobel named physics first in his will, I was the senior laureate; one of the most spine-tingling experiences of my life began as I stood just below and behind the elevated doorway at the rear of the platform. I had watched the award committee members and laureates from previous years climb the steps to that doorway and disappear. Now it was my turn. I stopped briefly in the doorway to allow the trumpeters to sound their flourish, then walked to my chair as the audience stood as one and applauded. The five other candidates followed in order: chemistry, medicine, and literature.

Much followed before the actual presentations, but not much registered permanently on my overstimulated brain. I do remember Brigit Nilsson singing an aria a few feet from my ear, a unique experience that I thoroughly enjoyed; my last time on such a stage had been as a spear carrier in *Aida.*

With the preliminaries out of the way, the Swedish Academy's Sten von Friesen went to the podium and told the king and the audience in Swedish what I had done that the committee considered worthy of a Nobel Prize. "Dr. Alvarez," he said to me then in English.

"Your contributions to physics are numerous and important. Today our attention is focused on the outstanding discoveries which you have made in the field of high-energy physics as a result of your far-sighted and bold development of the hydrogen bubble chamber into an instrument of great power and precision and of the means of handling and analyzing the large quantities of valuable information which it can produce." I stepped down from the platform to the throne on the main floor, received my diploma and medal from the king of Sweden, thanked him, and shook hands. Making my way back to the platform, I encountered Walt, Millie, and Jan in the front row; Jan and I touched hands as I went by. While the other candidates became laureates, I scanned the sea of look-alike penguins for my Berkeley colleagues, found them, and nodded when they waved. They had noticed me searching them out, and we shared the pleasure of making eye contact.

At the Nobel banquet that evening Jan was seated next to the king and enjoyed his company. We laureates were expected to speak briefly at the end of the banquet. Yasunari Kawabata's remarks were read in English translation. Lars Onsager, the Yale chemist, ended his comments in his native Norwegian, which was well received. Marshall Nirenberg spoke for the three medicine laureates. When my turn came to be called to the podium, I took along my wineglass hanging unnoticed in my left hand and set it on the podium's lower shelf. I acknowledged the debt I owed to Arthur Compton and to Ernest Lawrence, whose Radiation Laboratory had produced so many laureates. I acknowledged the work of my younger colleagues, which was implicitly recognized in my award. I mentioned Jan's important role in my life and finished by saying, "I know it's an old Swedish custom that a man must skoal his wife at a banquet under penalty of forfeit. So now, with your permission, I will skoal my Jan." I retrieved my glass then and toasted my wife, a Nobel banquet first, as the papers noted the next day.

I had one last and unexpected duty that evening, responding to the dinner speech the university students directed to the new laureates. The student who delivered the speech and I stood halfway up the great staircase that joined the two floors of the two large ballrooms. Finished then with official duties, I was free to enjoy myself. In the lower ballroom I danced with Jan and with all the Berkeley ladies. The party broke up about two in the morning. Just before the bell rang that transformed us from royalty back into pumpkins, we found a photographer who took our picture as a group.

THIRTEEN

Checking Up on Chephren

WHEN John F. Kennedy became President, in 1960, he appointed Jerome Wiesner of MIT chairman of the President's Science Advisory Committee, the PSAC, which Dwight Eisenhower had established in the wake of Sputnik. It was a powerful and nonpartisan body that served Presidents well until it was abolished by Richard Nixon in 1973, while I was a member.

Hearing of Jerry's appointment, I wrote him that I had reached a turning point in my career—Don Glaser had just received his solo Nobel Prize, and my difficulties with Ed McMillan were mounting—and he quickly appointed me a senior adviser to the Federal Aviation Administration task group that specified future air navigation and traffic control systems. Though the FAA was proposing that all aircraft should be equipped with radar transponders, on which I held the basic patent, my membership in the task group represented no conflict of interest; my patent wouldn't expire for several years, but I had assigned it long before to the government for the standard "one dollar plus other [nonexistent] valuable considerations."

I spent a great deal of time on Washington-based committees during the next few years. Ernest Lawrence had protected me from committee service on the Berkeley faculty, arguing that my work was more important and that I could make up for the omission later. I found no place later because I hadn't served in the trenches and knew few professors outside the physics department. From MIT and Los Alamos days, however, I was known to many members of Washington technical committees. They made room for me commensurate with my wartime and scientific accomplishments.

Jerry next appointed me to a committee chaired by Lloyd Berkner studying how the scientific community could help the nation improve its capabilities for fighting a limited, nonnuclear war. Not long after we started work, Lloyd suffered a heart attack and I became chairman. I hadn't visited the Pentagon since World War II; the only person I knew in the Department of Defense was its third-highest-ranking civilian, Harold Brown, a former Livermore director who was the Pentagon's director of defense research and engineering (and later, of course, secretary of defense). But doors open for chairmen of high-level committees, and things went smoothly.

Old physics friends on the limited-war committee included Norman Ramsey, now at Harvard, and William Shockley, the 1956 physics Nobel laureate. Military members all had flag rank, which compounded the committee's credibility. General Russell Volckmann was one of the more interesting. During the Japanese attack on the Philippines early in World War II, he had escaped from the Bataan death march and organized a guerrilla campaign. He recounted the story later in *We Remained* and, having mocked the United States's most famous five-star general's "I shall return," proceeded to retire. Other committee members came from think tanks like the Rand Corporation and the Institute for Defense Analysis, most notably Dan Ellsberg, who later exposed the secret Pentagon Papers.

Jan found summer work at the National Institutes of Health while I toured. Military experts interested in fighting old-fashioned wars briefed us at the Pentagon for a week. Then, traveling in a nonpressurized four-engine DC–4, we visited the Green Berets at Fort Bragg, North Carolina, and watched the Marines assault Puerto Rican beaches. At Guantánamo Bay we saw an impressive demonstration of low-altitude precision bombing. With two days' liberty we hopped over to Jamaica's Montego Bay, where I knew the owner of a luxurious hotel (as a frequent guest of the Loomises) and was able to negotiate rooms for everyone at low off-season rates. We had a wonderful time becoming better acquainted.

We flew on to visit the USS *Northampton,* stationed just outside the territorial waters of the Dominican Republic. The longtime dictator Rafael Trujillo had recently been assassinated, and that country was in turmoil. The *Northampton,* a specially outfitted communications ship, carried a senior foreign-service officer aboard and communicated directly with the Pentagon and the State Department. I had recently read one of C. F. Forester's Hornblower novels; Admiral Lord Hornblower's flagship had been stationed in that story close to the waters our destroyer patrolled. Since it took Hornblower weeks to communicate with London, he had to make his own decisions, never sure who was friend and who was foe. The *Northampton*'s senior officers talked directly with their Washington superiors over a scrambled phone channel and so effectively could make no decisions on their own.

After more briefings in Washington and a full range of immunizations, I led most of the committee on a tour of the Far East while Norm Ramsey and Bill Shockley took the rest to the Middle East. We were briefed on Oahu and happily gave up our DC–4 to fly first-class on Pan Am to Manila, where we visited Subic Bay and Clark Field. One of my college roommates, Cecil Combs, had been serving as a B–17 pilot based at Clark Field on December 8, 1941, when most of our aircraft were destroyed on the ground hours after the Pearl Harbor attack. Cece's experience was often on my mind during our visit; he survived to pursue a distinguished flying career.

I found I adjusted well to VIP treatment. As leader of our traveling circus, I was properly the first out the door to receive the base commander's welcome when the plane landed. His chauffeured limousine, flag flying from its fender, delivered me then to his headquarters, where briefings always started with a transparency announcing that the presentation had been "specially prepared for the Limited War Panel chaired by Dr. Luis W. Alvarez." An experienced colonel followed with a canned briefing about the base we were visiting. From other officers in other settings, we learned more informally what we had come thousands of miles to find out. And the real VIP in the Philippines was our guerrilla-fighting general, Russ Volckmann.

We flew to Saigon in a military plane. The only U.S. military in South Vietnam in the summer of 1961 were training personnel and logistics experts there to help the South Vietnamese armed forces learn to use U.S. equipment. But the civil war was ongoing. I called on Ambassador Fred Nolting the day we arrived; he had escaped assassination only a few days earlier while being driven from his home to the embassy. A man passing on a bicycle had dropped an armed hand grenade through the limousine's open window. Nolting watched the grenade roll under his legs and assumed he had only seconds to live. Fortunately for the ambassador the device fizzled.

That near-miss stayed fresh in my mind as the committee's jeep caravan threaded its way through the crowded Saigon streets or down some lonely jungle road. Behind the advance jeep with its swiveling machine gun, I occupied the backseat of the number one jeep in solitary splendor. I always offered to share my ride, and my offer was repeatedly and imaginatively declined.

We learned a great deal about how the war was being fought, but our briefings never took the measure of Ho Chi Minh's dedication to rid his country of all Western influence. We heard the South Vietnamese characterized as the good guys and the North Vietnamese as the bad, which demonstrates how little Americans understood of revolutions and revolutionaries. Revolutionaries are not always to be admired, but their dedication to principle does include putting their lives on the line, not merely asking foreigners to supply sophisticated weapons.

I was unsophisticated about revolutionary war and unwisely trusting. A Vietnamese physicist was then a visitor in our bubble chamber group. I found a note in the mailbox of my Saigon hotel asking me to call his cousin, who invited me to cocktails. A few days later the cousin drove me to his family's comfortable home. After a pleasant conversation he took me to a small restaurant. During the meal he asked if there was anything I hadn't seen in Vietnam that I would like to see. I said I hadn't seen a water buffalo. He said I was eating one. He offered to drive me outside the city to view a live water buffalo, and I accepted. It was still light when we left; we drove through the countryside on a traffic-free, first-class highway. In an hour I was back at my hotel.

The next morning the Vietnamese security officer assigned to me told me I had been stupid to drive into the country: I might have been delivered into captivity. He asked me how I knew who the man was and how the man knew I was in Saigon, since my physicist colleague couldn't have known. I didn't have answers to his questions. His soldiers had followed me, the security man told me; a kidnapping attempt would have landed me in the middle of a firefight. I've suspected from time to time that a guardian angel watched over me, but two cars full of soldiers armed with machine guns make an effective second line of defense.

After further briefings in Bangkok and a quick stopover in Hong Kong, we returned to the United States. Dan Ellsberg proposed we write our report away from the interruptions of Washington in Santa Monica, where Rand could supply secretarial help and secure our classified documents. When we had a draft in hand, I flew east to report to Jerry Wiesner.

I told him about our trip and our tentative conclusions in his second-floor PSAC office in the Old Executive Office Building overlooking the Rose Garden. After lunch in the White House basement dining room, he asked if I would like to meet the Boss. "Sure," I said. Jerry checked with several secretaries, and then we went through a hall and came upon President Kennedy studying a newspaper scrapbook of his latest activities. When Jerry introduced me, the President asked me if I was related to Dr. Walter Alvarez of the Mayo Clinic. Several years earlier, it seemed, Dad had invited him home for lunch after examining him at the clinic.

Jerry told the President that I had been in Vietnam, and Kennedy asked me how things were going there. I shaded the truth when I said that they seemed to be getting better. Most people feel compelled to tell the President what they think he'd like to hear. I was no exception; I'd have had difficulty saying that our policies were disastrous, that we should withdraw most of our advisers and avoid becoming mired in an Asian war. (I talked to Jack Kennedy again when he presented the Collier Trophy to the Mercury astronauts; I was a previous recipient for my invention of GCA and received my trophy from President Truman. I had a later appointment with President Kennedy to receive the National Medal of Science, an appointment abruptly canceled in Dallas. Lyndon Johnson did the honors later. Jack Kennedy was very much a hero of mine.)

One of the factors that permits guerrillas to defeat organized military units is that their familiarity with the local terrain allows them to operate under cover of darkness. If organized units could see at night better than the guerrillas they were fighting, I reasoned, then the imbalance between them might be corrected. In the report we prepared in Santa Monica, I emphasized that new technology was available that would enable the infantry to see nearly as well at night as during the day.

While my team at Berkeley had been developing bubble chambers of liquid hydrogen to visualize fast-particle tracks, George Reynolds at Princeton had been working on an alternative scheme using optical image amplifiers developed at the RCA Princeton Laboratory. An RCA consultant on the application of the image tube to astronomy, George had proposed using such tubes, which were considerably more sensitive than photographic film, to record charged-particle tracks passing through a fluorescent material like sodium iodide. I had quickly calculated that to collect enough light to see a track the system would need a very low f-number, which would reduce its depth of field below a useful range. In high-energy-physics experiments, furthermore, the ideal target is a proton; fluorescent materials, atomically more complex, were unsuitable.

I studied the RCA specification sheets. They indicated that devices could be built that would enable infantrymen to see clearly in starlight and for long distances in moonlight. We recommended that the Pentagon support this work at its night vision laboratory at Fort Belvoir, Virginia, which was then concentrating on infrared viewing devices of the World War II sniperscope variety. Postwar soldiers wouldn't turn on their infrared searchlights, for the same reason that submarine skippers wouldn't turn on their active sonar systems: it was too easy for an enemy to detect the intense radiated power long before enough of it could be reflected back to accomplish its purpose. So night vision devices and sonar systems are now almost exclusively passive, radiating no telltale power. When eventually we became directly involved in the Vietnam War, many of our soldiers were equipped with starlight scopes.

I rounded out my summer's work by briefing Secretary of Defense Robert McNamara and the Joint Chiefs of Staff in the Secretary's impressively large office. That was one of my most difficult speaking assignments; I was talking to a group of professional military officers, all of whom had spent their formative years fighting the greatest limited war of all times—World War II.

Jerry Wiesner next appointed me to the PSAC Limited War Panel and installed me as chairman of the PSAC Military Aircraft Panel. Our Aircraft Panel unfortunately was not able to affect policy. We recommended that the Pentagon reject the F–111; Defense Secretary Robert McNamara forced it down the Air Force's and the Navy's throats. We recommended that the Air Force buy converted Boeing 747's as its primary transport aircraft; instead it bought the outrageously expensive C–5, which may still hold the record for massive cost overruns.

I continued to chair the Aircraft Panel when its recommendations were regularly ignored because it gave me an unsurpassed view of my second-favorite field after physics and because the perks were outstanding. The Air Force and the Navy treated the panel, and its chairman, as figures of great authority. I don't know how they failed to learn that we were impotent, but they missed no opportunity to lobby us.

Lockheed decided that I should know more about the virtues of their F–104 Starfighter. For a time the F–104 had been the world's most exceptional jet fighter, the first to hold all three important records—maximum speed, maximum altitude, and minimum time to climb to operational altitude. Lockheed was still building F–104's and selling them throughout the world but the Air Force hadn't bought any recently. A fighter-pilot friend arranged a test flight for me in one of the few existing two-place F–104 trainers.

I was tested first in an altitude chamber to see if I could survive a high-altitude bailout. At the Lockheed Lancaster plant Jan and I inspected F–104's being assembled, then flew to a nearby Air Force base, where we had lunch with F–104 pilots and their wives. All the pilots swore the plane was the best jet fighter ever built. No one told Jan that in Europe the F–104 was known as the Widowmaker because of the large number of German air force crashes.

The F–104 had stubby six-foot wings; in those preastronaut days it was frequently called "the missile with the man in it." Since its engine's thrust was greater than the plane's weight, it could fly straight up. It was another elegant creation from Kelly Johnson's famous Skunk Works, which had also turned out the U–2 and the SR–71 in record time. Hordes of committees and inspectors routinely retard the design and manufacture of military aircraft. By contrast, Kelly Johnson was so much in demand that he was able to dictate his own terms. He wanted only approximate specifications, and he insisted on being left alone. In a year or less he would deliver a fine aircraft that exceeded all the specs.

We screamed down the runway with the afterburner on and then went almost straight up to sixty thousand feet. There we leveled off, and the pilot let me see for myself how responsive the plane was. We flew for an hour, corkscrewing in slow rolls through the sound barrier without noticeable vibration. Accelerating up to Mach 2, I had no sense of speed, but when the pilot suddenly cut off the afterburner it felt as if we had just smashed into a wall. We dropped down and flew through narrow Sierra valleys using the terrain-following radar developed to allow U.S. aircraft to penetrate the Soviet air defense system. Our landing was spectacular. As soon as we touched down, the pilot ejected a large parachute; billowing out behind, it quickly slowed us to braking speed. (Jan finally flew at Mach 2 when we came home from Europe on the Concorde; we left London at 10:30 A.M. and arrived in New York at 8:20 that same morning.)

The Navy's large congressional appropriations attest to the effectiveness of its public relations program. High-ranking military officers are pleasant people if you don't engage them in combat; they're selected in part for their social graces, which they need to impress congressmen. I've often wondered if such qualities are really desirable in battlefield commanders. To be fair, my profession's practice of choosing college teachers on the basis of their performance in the laboratory rather than in the classroom is equally debatable.

The Navy invited the members of the Aircraft Panel to spend a weekend on the carrier USS *Forrestal,* then steaming a hundred miles off Norfolk, Virginia. We flew out in a small Navy transport, seated facing backward to make more bearable the sudden deceleration when the plane's landing hook caught the carrier's arresting gear. On the *Forrestal* we talked with pilots as they rode the elevator with their planes up to the flight deck. We watched them catapulted down the launch runway and landing on the longer recovery runway. We saw bombing and strafing demonstrations against targets towed behind the carrier.

Intelligence officers in the pilots' ready room gave us a fascinating briefing. They asked us to pretend that we were carrier pilots in the Mediterranean when word came from Washington that we were at war with a particular Mediterranean country. Our job was to destroy enemy aircraft on the ground to prevent them from engaging us in aerial combat. I was surprised at the detailed information available on the military establishment of the designated enemy. Each of us was given precise maps indicating attack approach routes to fly to avoid radar detection. The Israeli pilots who attacked Egyptian bases in the 1967 Six-Day War must have had a similar briefing.

Much of the intelligence data came from photographic reconnaissance satellites, information I was not then cleared to receive. Later I became a member of the first review panel authorized to see all our satellite photographs plus the cameras that took the pictures. That assignment was much like my first exposure to the National Security Agency's code breaking, an exciting look at an important and highly secret enterprise about which I had known almost nothing. I've been told that the relevant satellite technology has been greatly improved in the decades since I reviewed it. But it's still governed by the laws of physical optics. With the best possible diffraction-limited camera at an altitude of 120 miles, two objects can't be closer together than a few inches and still be recognized as distinct. This resolution limit is adequate to distinguish aircraft or tank types and to count soldiers but not to read license plates, assuming we haven't yet orbited any optically perfect ten-foot mirrors.

As panel chairman, I was given the admiral's shore cabin on the *Forrestal,* a luxurious suite. While I was undressing for bed, there was a knock on the door. I opened it to Captain Pete Aurand, USN, President Kennedy's military aide. I had seen Pete only once since he had served as one of our GCA test pilots, but I had followed his distinguished career in the newspapers. We caught up on mutual friends; then he opened his briefcase and pitched the Navy's standard argument for building two more nuclear-powered carrier task forces. I wondered how the Navy had learned that Pete and I were old friends and why, in the light of the Aircraft Panel's demonstrated lack of influence in military affairs, they had transported him to the *Forrestal* at sea just to lobby me. They were certainly thorough.

The next morning Pete and I were photographed together watching the radar displays in the combat information center. Before we left the ship, each panel member was presented with an individualized scrapbook of photographs, including a photograph showing him sitting in the captain's chair on the bridge with the captain smiling alongside. Then we climbed back into our transport and were catapulted off the carrier.

Later I served on several National Aeronautics and Space Administration (NASA) committees. The most important to me personally was the Physics Advisory Committee that advised NASA's associate administrator, Homer Newell. I was having trouble finding support for my bubble chamber program, and Homer had extensive discretionary funds under his control. I proposed that he support a program to study high-energy cosmic rays as a temporary substitute for the extremely high-energy accelerators then under study but not yet built. He funded my group, which later became the Lawrence Berkeley Laboratory astrophysics program, at the rate of several million dollars a year. We flew superconducting magnets to high altitudes with balloons. It was an exciting program but not the most productive, and most of us went on to other studies. To commute from Berkeley to Chico, 250 miles away, where our balloon-flying operation was based, I was able to convince myself that I should own my own airplane, a nice bonus.

Among my other NASA-related committee activities, I was a longtime member of the Space Sciences Board, which has the statutory duty of advising the NASA administrator on scientific matters. I also served on the advisory committee to the director of the Apollo moon-landing program, which included among its perks watching Saturn launches from the VIP stands at Cape Canaveral and getting to know the astronauts in small, informal meetings. During 1985 and 1986 I was a presidential appointee on the National Commission on Space, a fifteen-person board charged with suggesting what the U.S. civilian space program should look like in the next fifty years. The two most famous members of our commission were Neil Armstrong and Chuck Yeager.

I was invited in 1964 to give the opening address at a meeting of the Australian–New Zealand Association for the Advancement of Science. Jan and I decided to take a side trip to New Zealand and to return via Tahiti. I also wanted to see Antarctica, then a strictly male domain to which the National Science Foundation controlled access with logistic support from the U.S. Navy. I stopped in at NSF on a visit to Washington and asked the Antarctic Program director if I could get over to the ice. He quickly obliged. So after Jan and I explored New Zealand's beautiful South Island, which included a ski plane landing on Mount Cook's Tasman Glacier, I reported to the Antarctic Program's Christchurch airport office. They booked me on the next Super Constellation and outfitted me with two large duffel bags filled with warm shoes, sweaters, jackets, mittens, and fur hats, which I was instructed to carry with me on all trips away from McMurdo.

The Antarctica map shows the Greenwich meridian continuing downward from the centrally located pole, thus defining West Antarctica to the left. The lower half of the map is called South Antarctica even though all Antarctica is north of the South Pole.

On the uneventful ten-hour flight directly south to McMurdo Sound, we flew for an hour along the Victoria Land coast with its magnificent snow-covered mountains and enormous glaciers. We landed on a long runway bulldozed out of the rough ice. Two GCA trucks serviced it. Their operators had heard about me in radar school and later invited me to inspect their equipment.

William Austin was my McMurdo host. With him I made a number of flights to outlying stations. Antarctica has two cultures, the scientists in bright orange jackets and the Navy in darker gear. Bill and the Navy commander had developed the fine personal relationship that one sees in military operations where lives are at risk. On my first day at McMurdo, Bill drove me in his snowcat to New Zealand's nearby Scott Base. When we were shown into the base commander's office, he stopped dictating to a beautiful blond secretary who turned out to be a realistic inflated balloon. Rows of huskies outside were tied to cables staked to the ice; they lived outdoors all summer. As I looked through the windows of the roughly built hut where Robert Scott spent the 1902 winter, I marveled at the hardiness of the early explorers. By contrast, I lived in a comfortable dormitory room with heavy black drapes to promote sleep—the sun remained well above the horizon twenty-four hours a day.

My first flights took me to Byrd and Eights stations, both dug below the surface of the ice. I never found anyone in charge of plane schedules, so I arranged other flights by buying drinks for flight crews in the officers' club. One was a fourteen-hour flight over mostly unexplored northern Antarctica to determine the condition of the ice for a surface party that would traverse it in tracked vehicles the following summer. I sat next to the Marine Corps navigator in the ski-equipped C–130; its sole cargo was a large auxiliary fuel tank. The navigator took sightings with a bubble octant from the plane's astrodome, since there were no useful radio beams and no observable stars. The sun's altitude only located us on a line; I was puzzled that the navigator was able to find our position. I finally concluded that the moon must be above the horizon and about ninety degrees to the left of the sun. (If it had been ninety degrees to the right, it would have been below the horizon.) When I looked out the window, there it was. That rare arrangement of sun and moon was of course what set the date for our exploratory flight. The sun and moon stayed at nearly the same elevation angles as they circled the horizon once each day.

The South Pole is located on level ice nine thousand feet above sea level, so one goes *up* to the pole. After passing the pole, we flew for hours at a constant barometric altitude and at a constant radar altitude of three hundred feet above the surface. It was hard for me to believe that ice, like water, seeks its own level to that accuracy. I helped locate possible crevasses on the surface as it passed rapidly and very close below us.

The Russian Vostok base is located at the pole of inaccessibility, the center of the largest circle that can be completely drawn within the continent. We flew over and then circled Vostok several times, but no one came out to greet us. In a U.S. military aircraft flying low over a Russian outpost, I was impressed that we were in no danger of being shot down. That impression reinforced my long-held belief that the utopian peace I envision between us and the Soviets can come to pass given enough time.

After Vostok we headed for the pole, where we were supposed to land on the ice and visit the Amundsen-Scott base. Unfortunately our fuel was so low that we circled the pole only twice (two quick round-the-world flights) and then flew over the vast and beautiful Beardmore Glacier, which carries the polar plateau ice hundreds of miles into McMurdo Sound. The next day my officers' club friends said they would make it up to me by getting me to the pole on a supply flight. I waited on the flight deck of a C–130 for six hours, but the flight was scrubbed because of bad weather at the pole.

That night at the club one of the pilots asked me if I would like to go over to Crozier, a small zoological research laboratory staffed with resident scientists, and see the penguins. I'd been hinting about such a treat for more than a week. Crozier, on the south side of Ross Island, is fifty miles from McMurdo. We flew in a small helicopter and circled Mount Erebus, an active volcano that makes up most of the island. The pilot had said the cabin was too crowded for my duffel bags, so I had left them behind for the first time.

When I got out of the helicopter to take movies, I told the pilot I'd be gone for only a few minutes. He assured me he wouldn't leave without me. As I was photographing some of the hundreds of thousands of penguins, though, the helicopter lifted off and flew away. I waved, sure it would return shortly to pick me up. When it didn't, I walked to the small research building and asked a zoologist when the helicopter would return. He said it would be back in a week. I was distressed. I didn't even have a toothbrush and was scheduled to meet Jan in Christchurch the next day. I explained my problem and learned that all scientists looked alike to the Navy. When two left and two climbed aboard, the pilot assumed I was one of them and took off. To radio the pilot on the other side of Mount Erebus, the message was relayed through a geological party on the Beardmore Glacier.

When the pilot returned, he wasn't happy. After a hearty dinner back in McMurdo with a visiting admiral, I flew to New Zealand in the cargo bay of a C–130, which was faster and more comfortable than the Super Constellation.

I wrote thank-you notes on Antarctica to some eighty people who had written to congratulate me on being the first physicist to win the National Medal of Science. My stationery pictured a penguin standing on the Crozier shore; the letters were flown up to the South Pole to be postmarked. I also collected a number of rocks that we later searched for magnetic monopoles.

I first saw the Egyptian pyramids in the summer of 1962 on my way from Victoria Falls to Geneva. They are extraordinarily large structures. The Great Pyramid of Cheops and the neighboring Second Pyramid of Chephren rise about 450 feet high and span more than 700 feet on a side. The Washington Monument, at 555.5 feet, was the first manmade structure taller than these pyramids, but it musters only 1 percent of their bulk and was built forty-five hundred years later. I tried to imagine how the ancient Egyptians constructed such massive monuments. A quick back-of-the-envelope calculation convinced me that the standard long-ramp scenario was untenable—the ramps would have been more massive than the pyramids.

I didn't find time to think hard about the pyramids and their building logistics until my visit to Antarctica. My calculations were exciting, and as soon as I returned home I checked out a tall stack of books on the subject from the university library.

I was struck by the cutaway views of the three large pyramids. Sneferu's had two chambers, his son Cheops's three, but no one had found any chambers at all in the huge bulk of his grandson Chephren's pyramid, the one that still preserves a cap of its smooth white limestone casing. When I learned that all the chambers had been discovered thousands of years after the pyramids were built, it occurred to me that there might be undiscovered chambers in the Second Pyramid of Chephren. The modern archaeological view, that pyramid architects had simply given up making chambers, was contrary to everything I knew of human nature. If Chephren's grandfather built two chambers and his father three, it seemed most likely to me that Chephren would have ordered four.

An improbable accident had disclosed the internal structure of Cheops's Great Pyramid. In the ninth century A.D. Mamun, the caliph of Cairo, decided that the pyramid guarded undiscovered treasure. He commissioned a tunneling team to search for it. With iron tools and battering rams, the tunnelers entered the Great Pyramid in the center of its north face, where all earlier pyramid entrances had been found. When they had driven their tunnel one hundred feet into the masonry, they heard a long, low rumbling beyond its east wall. Mamun ordered the tunnel turned eastward. They soon ran into an ascending passageway plugged with a granite block. All the internal structure of the Great Pyramid lies in a vertical plane displaced twenty-six feet east of the center of the pyramid, to fool grave robbers like Mamun. If the stone their battering had loosened had not tumbled down that displaced passageway while his team was there to hear it, they would probably have continued tunneling another 650 feet and emerged empty-handed at the center of the south face.

Digging around the granite block, crawling up the narrow ascending passageway, they soon reached the enormous sloping Grand Gallery, at the top of which was set the king's chamber. A horizontal passageway connected to the smaller queen's chamber directly below. The purposely displaced Great Pyramid chambers encouraged me to find a way to X-ray the Second Pyramid, certain that we would discover unlooted and even more cleverly hidden chambers.

As far as I know, the group I eventually assembled was the first to use cosmic rays for any practical purpose. We used cosmic-ray muons to probe Chephren's pyramid. Muons are ideal for probing an object that large but not much larger or smaller. Neutrons, X-rays, or gamma rays wouldn't have penetrated to our detectors. Neutrinos would have penetrated easily but wouldn't have given us any information, since to neutrinos rock and air are quite transparent. Unfortunately, there aren't any other such objects to probe, as I tell the prospectors who frequently ask for my help in locating lodes of gold ore.

I first discussed my project with a citizen of the United Arab Republic—Egypt—in 1964. I had been directing a summer school on Lake Como in Italy, and Jan and I drove over the Alps to visit friends in Geneva. I explained my ideas to Bogdan Maglich, who introduced me to a member of his research group, Fikhry Hassan. Soon afterward Bogdan wrote me that Fathy El Bedewi, the head of the physics department at Ein Shams University in Cairo, was interested in collaborating. Bedewi and I then began a voluminous correspondence that led to the Joint Pyramid Project of the United Arab Republic and the United States of America. Bedewi enlisted the support of archaeologists and government leaders in his country, and I did the same in mine. At a 1965 international atomic-energy conference in Tokyo, he discussed our proposed search with Glenn Seaborg, then chairman of the AEC, and persuaded Glenn to return to Washington via Cairo.

In the meantime I made contact with the Egyptian Department of Antiquities, which administers the pyramids. I had just finished reading *The Pyramids,* by the great Egyptologist Ahmed Fakhry, and had concluded that he was exactly the person we needed for our embryonic project. I called the only archaeologist I knew, the University of Chicago's John Wilson, and asked him how to get in touch with Fakhry. John suggested I call a Berkeley hotel; he had boarded Fakhry on a California-bound plane only the day before. I was soon talking with the Fakhrys in their hotel room. We liked each other immediately. He joined the project, becoming an ambassador for us between the two-nation scientific team and the archaeological community.

Back in Washington, Glenn Seaborg persuaded his AEC colleagues to support us. Then we officially established the Joint Pyramid Project, its establishing document executed in Cairo by diplomatic representatives of our two countries. Its administration was officially entrusted to a committee of distinguished Egyptian scholars. A three-man executive committee—Bedewi, Fakhry, and I—managed day-to-day operations. The Smithsonian Institution supported our project with Egyptian pounds owed for food the United States had supplied Egypt, which under Public Law 480 could be spent only in Egypt for scholarly purposes. We discovered that TWA would accept Egyptian pounds in payment for air fares, which permitted the key Joint Pyramid Project people to make frequent visits.

Word of my exciting project spread through the Radiation Laboratory. As with most of my projects, the people who played key roles voted with their feet and joined before the funds arrived. Jerry Anderson, who discovered the rho meson while working for his Ph.D. as a member of my bubble chamber group, emerged as the leader of our experimental team.

We invited four Ein Shams faculty members to Berkeley to participate in the meetings at which we designed our detection equipment. It would detect more or fewer muons, depending on the thickness of masonry they had to traverse to arrive at the point of detection, the Belzoni Chamber, located under the Second Pyramid just north of the center. Muon intensity should increase in the direction of a hidden chamber within the main mass, since muons arriving from that direction would have passed through less rock. To measure the muons, we would install two spark chambers each six feet square that sparked when charged particles passed through them. Scintillation counters mounted above and below the spark chambers would trigger the chambers when both counters signaled the passage of a penetrating muon. Digital circuits would count the muons and plot their angle of arrival.

By September 1966 Jerry Anderson, my son Walt, and I, with our wives, were ready to go to Cairo. Jerry brought equipment mock-ups to make sure everything would fit through the pyramid passages. Ahmed Fakhry arranged for the Joint Pyramid Project to occupy a modern building near the Second Pyramid's Belzoni entrance. It had been his home and office when he was chief inspector of pyramids. We converted it into a combined office and shop complete with electronic instrumentation and machine tools. I have long been a member of the board of directors of Hewlett-Packard. Because of Bill Hewlett's interest in the Joint Pyramid Project, Hewlett-Packard made a gift of the electronic equipment. The National Geographic Society contributed substantial unrestricted funds, as did my friend Bill Golden. IBM donated a computer, the first any Egyptian university owned.

By the end of 1966 the equipment had been built, tested, and installed in the Belzoni Chamber. It was joined early in 1967 by six members of the Berkeley contingent and their wives. Most shipped their personal possessions; we expected the project to last two years.

The equipment first operated on June 5. The next day the Egyptian-Israeli Six-Day War broke out. Early that June morning Israeli aircraft successfully avoided Egyptian radar and destroyed almost the entire Egyptian air force on the ground. The Egyptians were as outraged as Americans had been when Pearl Harbor was attacked. Egypt accused the United States of helping Israel, broke off diplomatic relations, and interned our foreign-service officers.

Jerry Anderson and his colleagues wandered downtown Cairo that afternoon. Jerry took movies of the crowds until the police noticed him, seized his camera, and confiscated his film. The Americans escaped the hostile crowds by retreating to the U.S. embassy. From there they were sent by train to Alexandria. Jerry was seized getting off the train, detained, and hostilely interrogated. After three hours he was released. The next day another member of the group, Fred

Kreiss, was hauled off and briefly questioned. In due time a chartered ship carried the Americans to safety in Athens.

One of the questions his interrogator asked Fred triggered an interesting fantasy. What were they *really* doing in Egypt? Fred was asked. I imagined myself an Egyptian intelligence officer trying to understand why Egyptian radar had failed to detect the incoming Israeli planes. In that role I remembered hearing about mysterious electronic equipment installed in the Second Pyramid, the finest of bomb shelters. The equipment was reported to have been put into operation on the day before the raid. I set to work to look into this fellow Alvarez. I discover that although he says he is doing archaeology he has published no papers whatsoever in that field. Then I discover that he has patented an electronic system, VIXEN, that foxes radar systems so that their operators don't know an aircraft is approaching until it's less than a hundred yards away. I conclude that the diabolical Alvarez is a threat to my country and that the United States colluded with Israel to destroy the Egyptian air force.

After the war we in the project were among the first Americans to return to Egypt. We were welcomed with open arms, indicating that my fantasy was just that. Our equipment was in good shape, and we went back to work.

In February 1968 Lauren Yazolino replaced Jerry in Cairo. Enthusiastic letters from Fakhry and Bedewi told us that measurements were progressing well. Hundreds of thousands of cosmic rays had been tracked down through the pyramid by our two layers of spark chambers. The data tapes served as the basis for a scatter plot, one point for each cosmic ray. This plot was the equivalent of an enormous horizontal X-ray film resting on the point of the pyramid with the X-ray source located in the Belzoni Chamber. The pyramid's diagonal ridges and its distinctive limestone cap were clearly visible because fewer muons could penetrate the thicker rock in their directions.

More muons would cause extra blackening by penetrating through the lesser density of the rooms we were convinced must be present. As more muons accumulated, our plot's graininess smoothed out. Watching them accumulate was like watching a photograph developing in a dimly lit darkroom: the main outlines show up first; then the finer details appear.

Two months after I'd returned to Berkeley from working with Lauren, he sent me a telex. "We have found an EGG," it said. "Histograms show much more than we expected." Lauren assumed I would understand EGG to mean "East Grand Gallery." I didn't; I was too busy packing. I already had a visa. I kissed Jan good-bye, and off I went.

Tutankhamen was a minor monarch who reigned for only three years and died still a boy. The contents of his tomb were such that its discovery by Howard Carter and Lord Carnarvon is still considered the greatest triumph of archaeology. But Chephren was a great pharaoh, and he reigned long enough to build a great pyramid. As soon as I received Lauren's cable, I began dreaming of the golden artifacts we would see as we looked into Chephren's tomb, intact for forty-five hundred years, for the first time.

The Department of Antiquities had agreed that, if we found a cavity, we could drill in that direction until we broke into the chamber and could then introduce a small camera and take flash photographs. If the photographs showed "wonderful things"—Carter's first words when he looked into the tomb of King Tut—a horizontal tunnel would be dug into the chamber from the pyramid face. I can't express the exhilaration I felt at the prospect of making the greatest discovery in the history of archaeology. When I read Carter's description of his "wonderful things," a chill had run up my back. So in my Walter Mitty–like dreams I pictured myself standing behind my dear friend Ahmed Fakhry when he first peered into Chephren's burial chamber and described what he saw. My work in physics might soon be forgotten, but I would go down in history as a great archaeologist!

When I arrived at the Ein Shams Computer Center, Lauren and Bedewi showed me bar charts of muon counts plotted against compass direction and elevation angle. The plot was much more dramatic than I had expected. Too much so: the apparent chamber let in twice the number of muons that penetrated the pyramid in neighboring directions. A chamber that big would certainly cause the pyramid to collapse.

We found our error, a programming bug. As the team on site uncovered it, I kept on asking questions. If they had found it themselves, I would have missed two of the most exciting days of my life. Besides the real phenomena you uncover in research, you find many other apparent phenomena that don't survive careful examination. The disappointment of learning that a discovery is false hardly subtracts at all from the elation you feel while you believe it to be real. I've experienced this powerful but unrealistic elation many times.

Gerry Lynch finished our data analysis in time for me to give talks about our results in Sweden after I received my Nobel Prize. The Second Pyramid unexpectedly turned out to be quite solid. The data completely replicated the pyramid down to the details of its irregular cap. At a meeting of the American Physical Society in April 1969 I reported that in the 19 percent of the structure we searched—a 35-degree half-angle cone above the Belzoni Chamber—there were no chambers.

Our increased respect for the clever Old Kingdom architects' abilities to thwart potential grave robbers, a class that now included us, made us anxious to finish probing the whole pyramid. This we managed to do over the next several years with rebuilt equipment that could be tilted and rotated in every compass direction. The pyramid was solid throughout.

"I hear you didn't find a chamber," people frequently say to me. "It wasn't that we didn't find a chamber," I reply. "We found that there wasn't any chamber." But even that statement isn't quite correct. Gerry Lynch, analyzing the data from the second round of measurements, discovered a large chamber, the Belzoni Chamber itself. He hadn't put the Belzoni Chamber into his computer model of the Second Pyramid, the standard against which we compared our muon plots. The observed number of counts from low in the west was greater than the calculated number because the detector was set against the eastern wall, which gave the western arrivals a free ride through thirty feet of air. We didn't see this effect in the scatter plots, because the human eye isn't sensitive to slow variations of intensity with direction, and we hadn't seen it in the original experiment, because we were then looking upward. The fact that Belzoni had discovered the underground chamber first, 154 years before we did, doesn't diminish the importance of Gerry's proof that cosmic rays could do the job. If Chephren had commanded a chamber to be built in his pyramid, we would have found it.

I last visited Cairo in 1979, several years after the Joint Pyramid Project ended, to help celebrate the Ein Shams University Computer Center's tenth anniversary. The Faculty of Science awarded me an honorary doctor of science degree; that evening, at dinner in Bedewi's home, we watched the noontime White House signing of the Camp David accords. Anwar Sadat embraced Menachem Begin while President Jimmy Carter looked on with obvious pleasure. I couldn't imagine a more fitting climax to our project, which had fostered friendship and cooperation between the scientists and scholars of our two countries while our governments refused to talk to each other.

FOURTEEN

Scientific Detective Work

AS a graduate student, I heard the renowned American physicist R. W. Wood lecture on scientific detective work. Wood was frequently called on by the Baltimore police to help with difficult cases. In his lecture he described the case of a woman who died mysteriously while sitting alone in front of her fireplace. The coroner found a tiny metal fragment in her heart and a barely visible entrance wound in her chest. By spectroscopic analysis Wood showed that the metal involved was a copper alloy used exclusively in blasting caps. Apparently the cap had been embedded in the firewood and the heat of the fire had detonated it, propelling a fragment of its casing into the woman's chest at a velocity much higher than that of most bullets. I was to learn in the course of time that much of a physicist's life, both in and out of his laboratory, is spent in scientific detective work.

Wood's best-known work of detection was the discrediting of N-rays, a notorious early-twentieth-century will-o'-the-wisp. The British had discovered cathode rays, the Germans X-rays, the French radioactivity. The search for rays was a scientific and nationalistic olympiad. In 1903 Professor Blondlot of the University of Nancy announced his discovery of N-rays, which he claimed were emitted by a strange collection of subjects and absorbed by an equally diverse group of objects. Hundreds of N-ray papers followed. The French frequently confirmed Blondlot's results; the British and the Germans usually found them unreproducible.

Wood visited Blondlot's laboratory, where a cordial demonstration of prism-refracted N-rays was provided. A simple spectrometer in a dark room measured the refraction angles of the strongest N-ray lines. As evidence for N-rays Blondlot used the intensity change of a weak luminescent spot. He set the spectrometer on the peak intensity of a line, and his assistant in an adjacent lighted room recorded the angle. The settings were repeatable run to run.

Before the final run, however, Wood surreptitiously removed the prism. Blondlot nonetheless called out the N-ray line positions that agreed with the earlier runs. Wood replaced the prism before the lights went on and then behaved in an inexcusably uncivilized way: he didn't tell Blondlot what he had done. He might have coauthored a paper with Blondlot in which the French scientist retracted his earlier claims as the result of a "definitive test" that Wood had proposed. That would have been scientifically acceptable. Then everyone would still remember Wood's key role in demonstrating that N-rays didn't exist, but Blondlot would have saved face. Instead Wood left Blondlot's laboratory without revealing his secret and wrote a letter to *Nature* exposing the whole sad story. Blondlot lived out his life under a cloud, and Wood, by holding a fellow scientist up to public ridicule, very probably forfeited his chance for the Nobel Prize he might otherwise have won.

I shot down the claim of a Berkeley colleague to have discovered a magnetic monopole in cosmic rays, but much more politely. The magnetic monopole is a hypothetical entity, the source of a magnetic field analogous to the electric field of an electric charge (a negative field for the electron, a positive field for the proton). All known magnets are made up of magnetic dipoles; if a bar magnet is cut in half, the two halves will each show north and south polarities; so will progressively smaller cuttings. I have searched extensively and fruitlessly for magnetic monopoles in such places as Fermilab proton interactions and the dust the Apollo astronauts brought back from the moon.

My search of moon dust was criticized in the pages of *Nature.* In a letter I wrote to the editor partly quoting my original proposal to NASA, I explained why such a search was worthwhile:

> "Modern experimental physicists are generally agreed that the discovery of any one of the following would constitute a major breakthrough in our

understanding of the physical universe: (1) free magnetic monopoles, (2) the heavy particles (quarks, etc.) whose existence has been postulated to explain the observed regularities in the 'spectroscopy of fundamental particles,' and (3) gravitons, or some observation of gravity waves."

Searches for physical phenomena of the kind just mentioned have a special fascination for experimental physicists in that if one multiplies the chance for success (which is admittedly very low) by the scientific importance (which is enormous), the product compares favorably with that of the more routine experiments on which one spends most of one's scientific life.

That my Berkeley colleague Buford Price thought he had discovered the magnetic monopole was something I learned when I arrived at my office one day in 1975 and found him going through my extensive file of reprints of monopole studies, to which my secretary had quite properly given him access. Buford showed me his impressive evidence, and I congratulated him. He soon announced his discovery informally at physics department and laboratory talks. On both occasions I presented older evidence in disagreement with his observation but didn't try to rule his discovery out.

He soon left for a cosmic-ray conference in Munich, where he planned to announce his discovery formally. I spent much of the intervening time learning all I could about the powerful techniques he had helped pioneer and had used in his experiment. Buford caught his supposed monopole in a carefully designed stack of plastic sheets, nuclear emulsion, and supersensitive photographic film. A monopole should be heavily ionizing, and the stack of sheets was appropriately etched by incoming ionizing particles. A special photographic film designed to record Cerenkov radiation, a sort of charged-particle sonic boom, was placed near the top of the plastic stack, with a slab of nuclear emulsion just below it.

I eventually succeeded in finding a more conventional explanation for Buford's observations. In any scientific discovery, and especially in what might be a great discovery, a true first, it's absolutely essential to rule out all other alternatives. Buford hadn't done so. What he thought was a magnetic monopole was more likely to have been a platinum nucleus fragmenting as it passed through his package of plastic sheets, first to osmium and then to tantalum. A key test concerned the thickness of the package. If it was as thick as Buford indicated in his paper, my alternative explanation wouldn't have worked according to my scenario. I called his Houston collaborators to ask whether the package wasn't really less than half the quoted thickness; the answer was yes. I had never before had occasion to question the value of such a simple measurement in a scientific paper. But this error confirmed convincingly the correctness of my explanation.

I had Wood and N-rays in mind as I wrestled with the delicate question of what to do with my new information. An international conference on high-energy physics was under way at Stanford at the time at which I finally decided to present my findings. By now I had learned, at third hand and then, in a long telephone call, at first hand that Peter Fowler, Ernest Rutherford's grandson and the acknowledged world expert on the use of nuclear emulsions in heavy-ion physics, had independently arrived at the same platinum-to-osmium-to-tantalum sequences of events. He was about to become Buford's Munich roommate and would attempt to dissuade him from announcing his supposed discovery and, failing that, would publicly challenge Buford's evidence there.

Before contacting Stanford, I first called Buford and suggested that we coauthor a paper interpreting his event as a cosmic-ray platinum nucleus fragmenting twice on its way through his stack. If he accepted, I wouldn't give my paper. I knew he would be terribly disappointed, as I had been on several occasions when my discoveries evaporated minutes after I was sure they were certain. But I felt he would be relieved not to have to retract his claim later. We talked for a solid hour, but I was unable to convince him. He didn't believe that he had made a substantive error in recording his package thickness. He would check it, he said, and if I was correct he would cable the conference chairman.

I called one of the Stanford conference organizers then and told him I had a paper I would like to present as a postdeadline contribution to the last session. He asked what I was going to report. I said that since I hadn't learned anything new at a conference in thirty years I wanted the privilege Wilhelm Roentgen had enjoyed when he surprised his colleagues with news of X-rays without preparing them in advance. My contact declined to schedule my paper blind, but I tracked down another organizer, Samuel Berman, who immediately understood and scheduled my talk for the last period before lunch the next day.

Jan and I waited in the front row while the auditorium at Stanford filled to overflowing. The word was out that I had something interesting to say on an unknown subject. My talk examined the evidence for a magnetic monopole and the alternatives thoroughly; it was, I think, a good example of correct, serious scientific criticism. I ended it by thanking Buford for "his complete openness and obvious desire to have all the facts of the case made known." The session chairman then read Buford's cable confirming my new information on the thickness of his film package. After my talk and Fowler's in Munich, no one accepted the Price event as evidence for a magnetic monopole. As I left the auditorium and walked into the cafeteria, Richard Feynman moved from his table to meet me. Dick used to crack the safes at Los Alamos where we stored top-secret documents, as he recounts in his humorous and best-selling autobiography. "That was a great detective job!" he told me. No physicist could ask for higher praise.

The scientific detective work for which I am best known concerned the assassination of President Kennedy. Three years after that tragic event *Life* published the first color enlargements of frames from the Abraham Zapruder eight-millimeter film that the Warren Commission investigating the assassination had considered important evidence. From my years of experience in analyzing bubble chamber film, I was impressed with the wealth of information stored in those images, and I spent the Thanksgiving weekend analyzing them in detail. The Zapruder film, bought by *Life* before it was developed, provided a continuous telephoto record of the President's actions from his approach to the Texas Book Depository, where Lee Harvey Oswald was employed, to the railroad overpass where the motorcade disappeared from sight. The President's head was hit by the fatal shot in frame 313, when Zapruder was filming about fifty feet away from the open limousine.

My first observation and conclusions turned out to be wrong. I've had that experience often. But by the time I realized they were wrong I had spent so many hours looking at the photographs that I saw other, unexpected evidence for which I soon found an explanation that was correct and valuable.

I shared my new knowledge first of all with my friend and lawyer Edwin Huddleson. Ed was aware of the controversies surrounding the Warren Report, as I was not. He recognized that my observations, which pinpointed the timing of the shots fired at the President, were newsworthy. He called his law school classmate Richard Salant, then president of CBS News. CBS was planning a program on the questions raised by the Warren Report. Salant decided that my information could tie the program together. As a result it was first presented publicly in the course of a four-part documentary hosted by Walter Cronkite beginning in May 1967. (A full report was later published in the *American Journal of Physics.*) Interviewed in my living room, I explained how I determined from the Zapruder film when the shots were fired.

The spots where sunlight reflected from the body of the presidential limousine, I observed, were drawn out into parallel lines of equal length in some frames. These streaks indicated that the angular velocity of the photographer's camera had not matched that of his subject, so that the film image moved slightly during the thirty milliseconds when the shutter was open. Zapruder kept the car well centered in his field of view most of the time, indicating that he was an excellent tracker. The streak lengths varied significantly in different frames. I plotted their frame-to-frame differences, which were proportional to the angular acceleration of the camera's axis and a measure of sudden torques applied to the camera. Newton's second law says that if no forces were applied, the relative velocity of the camera axis and automobile would be constant, as would the streak lengths.

When I made my discovery over the Thanksgiving weekend after the *Life* photographs arrived, I had to wait until Monday morning to borrow one of the many Warren Report volumes from the law school library, measure all the streak lengths, and plot their differences against frame number. My graph showed streak episodes that I attributed to Zapruder's involuntary neuromuscular responses to the sounds of shots. In a CBS reenactment three men with identical movie cameras were stationed at the same distance from the road as Zapruder had been and told to track a car moving as the President's had moved. Without warning, CBS had shots fired from a tower positioned to simulate Oswald's location in the Texas Book Depository. The resulting films showed streaks similar to the streaks I had studied. CBS concurred that the timing of the shots fired at the President could be determined by my analysis.

There was no doubt when the shot that killed the President was fired, since his head disintegrated in frame 313. There were doubts about the timing of two earlier shots. The President clutched his throat in one frame; some took that act to mark the first shot. Others, including the members of the Warren Commission, believed Oswald had shot at the President a few seconds earlier but had missed because his view was partly obscured by a tree. I showed that the first shot had indeed missed and that the shot in the throat was the second.

I was unfamiliar then with the crowd of assassination buffs who believed that conspiracy was afoot and that the Warren Commission conclusions were a cover-up. Some were convinced that both Oswald and a second gunman shot the President; others thought Oswald wasn't even involved. I have found the buffs' books justifying their particular versions of conspiracy both unconvincing and incredibly dull. Most of them are mutually inconsistent. A single theme ties them together—that those in power are congenital liars, as is supposedly demonstrated further in Vietnam and at Watergate. And they appear to have accomplished their purpose; opinion polls indicate most Americans still believe that Oswald didn't act alone and that the Warren Commission covered up a conspiracy.

Paul Hoch, who received his Ph.D. in my bubble chamber group, has been my main contact with the buffs. He's devoted all his spare time for many years to buff activities, including publication of an exhaustively researched and well-written news bulletin, *Echoes of Conspiracy,* issued several times a year.

Paul tried for some time to interest me in the backward snap of the President's head immediately after frame 313. At first I thought that the President's head had fallen backward when the muscles of his neck no longer received instructions from his dying brain to hold it erect. Paul disagreed and insisted that the snap proved the President had been hit simultaneously by bullets from the front and the back, thus confirming the second-gunman theories. I thought Paul's argument was too silly to take seriously until he handed me a book, Josiah Thompson's *Six Seconds in Dallas,* when I was leaving for a meeting in St. Louis. I had time to waste on the flight, so I read the book, which included excellent photographs and carefully prepared graphs. When I studied the graph showing the changing position of the President's head relative to the moving car's coordinate system, I was finally convinced that the assassination buffs were right; there had to be a real explanation of the fact that the President's head did not fall but was driven back by some real force.

I then had to find an explanation for that force. Since I knew more physics than the buffs, it didn't take me long. The answer turned out to be simpler than I expected. I solved the problem to my own satisfaction on the back of an envelope.

I concluded that the retrograde motion of the President's head, in response to the rifle bullet shot, is consistent with the law of conservation of momentum if one pays attention to the law of conservation of energy as well and includes the momentum of *all* the material in the problem. The simplest way to see where I differed from most of the critics is to note that they treated the problem as though it involved only two interacting masses, the bullet and the head. My analysis involved three interacting masses: the bullet, the head, and the jet of brain matter observable in frame 313. The jet can carry forward more momentum than was brought in by the bullet. As a result the head recoils backward as a rocket recoils when its fuel is ejected.

I found it painful to consider in detail what happened to my hero John Kennedy, and thought of the problem instead as an abstract experiment involving melons reinforced with tape. When a bullet enters a melon it imparts some of its kinetic energy to the melon pulp, driving it out through the larger hole it makes as it exits. Momentum is equal to the square root of twice the kinetic energy times the mass of the material with that energy. If the forward-directed mass of melon pulp has one hundred times the mass of the bullet, the pulp-jet momentum is three times that of the entering bullet (assuming only 10 percent of the bullet's original kinetic energy is transferred to the jet). So the incoming bullet in effect activates a small rocket engine inside the melon that ejects its high-speed fuel in the forward direction, driving the melon, ten times heavier, backward with about twice the speed it would have moved forward if it had been solid. My order-of-magnitude calculations agreed with measurements made on the Zapruder frames immediately following 313.

Back in Berkeley I showed my calculations to Paul. He said no one would believe my conclusions, himself included, unless we could demonstrate the retrograde recoil on a firing range using a reasonable facsimile of a human head as a target. I discussed my theory with my longtime friend and Radiation Laboratory associate, Sharon ("Buck") Buckingham. Buck was an enthusiastic deer hunter; he offered his services, provided I paid for the melons into which he would fire the shots.

Buck carried out his experiments in June 1969, at the San Leandro municipal firing range, using melons wrapped with glass-fiber tape. Before Buck began firing, the expert marksmen in attendance told him he was wasting his time. "I've been around guns all my life," one said, "and you must be out of your mind to believe something you hit with a bullet will come back at you." The first test firings showed retrograde recoil; most of the reinforced melons were driven toward Buck, as I expected, rather than away.

Paul then said he wouldn't ask his fellow buffs to believe the test results unless he had photographic evidence to document the case. He enlisted the help of another physics graduate student and assassination buff who owned a remote-controlled Super 8 movie camera. I was present as an observer as well. We were all impressed to find that Buck's early results could be duplicated before the camera. The performances were now more uniform; six out of seven reinforced melons clearly recoiled toward the gun. These conclusive experiments, needless to say, failed to convince the second-gun theorists.

Besides the question of head motion, I was able to resolve several other questions as well. An eight-millimeter movie camera operates at eighteen frames per second normal speed and at forty-eight frames per second slow motion. Multiple-gunmen theories made more sense if the Zapruder camera had been running in slow motion, spacing the three shots across too short an interval to allow Oswald time to fire them all. We recognize a slow-motion "instant replay" sequence in a televised football game because we know how fast the players move and fall under gravity. Similar clues were apparently absent in the Zapruder film; it showed an automobile in which the occupants were sitting still, two motorcycle policemen immobile on their saddles, and a background of fixed structures and spectators standing still.

After many hours of studying the film, I found a "clock." In frames 278 to 297 a man applauded the President. I measured the space between his clapping hands as a function of the frame number and calculated that the needed muscle power at constant amplitude varies as the cube of the clapping frequency. Clapping in time with a metronome, I found that the spectator in Dealey Plaza was clapping at nearly the maximum rate consistent with his measured maximum hand spacing—provided that the film was running at eighteen frames per second. If the film had been running at forty-eight frames per second, the man would have had to use nineteen times more muscle power, a clear impossibility. The finding and interpreting of my cube-law clock showed that I was now able to relate physics to commonplace events as I had not yet been in the years when gymnastics occupied so much of my time.

The FBI expert who examined the Zapruder film testified before the Warren Commission that he could not pinpoint the precise location of the President's limousine during the important period from three seconds before to one and a half seconds after the fatal shot, because the background, a grassy plot, showed no reference objects like poles or buildings, only people walking.

I was able to pinpoint the car's location to within a few inches in each of the 160 frames that included this interval. I found sunlight reflected from a small shiny object lying in the grass in 21 of these frames. Equally important were the people walking in the background. The FBI photointerpreter didn't realize that a person whose weight is shifted solidly onto one leg is a reliable, if temporary, bench mark.

When I plotted the position of the limousine during this period, I discovered that it moved steadily at twelve miles per hour until frame 300, when it suddenly slowed down to eight miles per hour just before the President was killed and for a full second thereafter. The driver incorrectly but understandably testified that upon hearing shots he remembered speeding up. He did eventually speed up, and that was what he recalled. But the evidence is incontrovertible that he slowed down first. I worried about this discrepancy for some time. Then I found testimony that a police siren had sounded immediately after the President was shot. The many inconsistencies in the various witnesses' memories made me feel that it was permissible to suggest that the siren, from an escorting police vehicle behind the President's limousine, had sounded a few seconds *before* the fatal shot, but after the second shot. It would be most probable that an escorting officer, having heard one shot and seeing the President wounded by a second shot, would hit his siren button. The driver then instinctively took his foot off the accelerator. Unfortunately, his action made the President's head an even easier target for Oswald to hit. But by that time the angular velocity of the President's head as seen from Oswald's window was so small that even at twelve miles an hour the assassin could hardly have missed.

My second involvement in the Kennedy assassination started with a phone call from Phil Handler, the president of the National Academy of Sciences. Two years earlier the U.S. House of Representatives' Subcommittee on Assassinations had concluded that acoustical evidence established that two gunmen fired at the President. Their conclusion was based on testimony from two groups of acoustics experts with excellent credentials who examined audio records made by the Dallas police department. They believed that the sounds on the records, picked up by an open microphone on a policeman's motorcycle, included gunfire, and claimed to identify a fourth shot from the grassy knoll across from the motorcade by its characteristic echo reverberating from the surrounding buildings.

The U.S. Department of Justice had asked the academy to set up a committee to review this evidence, and Phil in turn asked me to chair it. Since the buffs would automatically have rejected any report published under my name, I agreed to be a committee member but suggested Norman Ramsey as a competent and acceptable chairman.

Our twelve-member Committee on Ballistic Acoustics embarked then on a fascinating eighteen-month study. I played an active role in the detective work. We were congenial individuals with good credentials in fields that together spanned much of the scientific spectrum, and we produced a report superior to anything that any of us had anticipated.

I found the testimony of the acoustics experts in the House subcommittee report amateurish, a view reaffirmed in a more careful review and in which the other academy committee members concurred. At our first meeting the two acoustics groups reviewed their testimony and answered our questions. With one expert I was distressed by my inability to get a simple point across. One of his illustrations showed that many of the recorded echoes in the Dealey Plaza reenactment of the shooting had reached the microphone array as concave wave fronts. Even a high school physics student would know that almost all such wave fronts must be convex, but the expert refused to agree that anything was wrong with his drawing. It was a new experience for me to watch a Ph.D. physicist stonewall in a technical argument. Some hours later he backed down, claiming that he hadn't understood my point and that a draftsman had connected some unrelated points with lines, a minor error that had escaped his attention and in no way influenced his conclusions. I realized then that a professional expert witness can't afford to admit that he has made a mistake. Otherwise he'll be confronted with a rereading of his admission in every subsequent trial. In this case, however, the setting wasn't forensic; the man was facing a jury of his peers, who frequently make mistakes and who look with scorn on the rare scientist who doesn't admit a mistake when it is pointed out.

Steve Barber, an Ohio buff who was also an accomplished musician, pointed our committee in the right direction. He wrote us that he could hear cross talk on the tapes between police channel II and police channel I. The presidential motorcade used channel II; channel I was the normal police frequency. The motorcycle (reputed to be) in the motorcade with the open mike was tuned for some reason to channel I rather than to channel II. Barber detected the order "Hold everything secure" on the open-mike tape in coincidence with the supposed shots. The police chief, he pointed out, had spoken those words on channel II fifty seconds after saying "Go to the hospital," *after* the President was shot. If we could prove that the "Hold everything" sequence on channel II had been broadcast by a loudspeaker on a nearby motorcycle and overheard by the open mike, then we could prove that the supposed shots were not related to the assassination.

We had access to the finest sound-recording and -analyzing equipment; with it members of our committee skilled in acoustics made sound spectra of the "Hold everything" sequence from both channels. The analysis of such spectra, like the analysis of handwriting, is ordinarily subjective, experts often disagreeing. We devised a new way of analyzing sound spectra that was completely objective, proving not only that the two recorded phrases were uttered by the same person but also—our innovation—at the same time on both channels.

The most conclusive part of the analysis was the work of Richard Garwin and two IBM colleagues. They used computer techniques to show that other features of the two records were substantively different, and for understandable reasons, proving that one wasn't merely a copy of the other. The two motorcycles were close together, but their radios were tuned to different frequencies. One picked up the sounds from the other. The inescapable conclusion was that the noises identified as shots by the "experts" had been recorded almost a full minute after the President had been hit.

Dick Garwin and I had previously served as members of many committees involved in matters of national security. One was an Air Force committee looking into a report from the Defense Intelligence Agency (DIA) that South Africa, with the possible assistance of Israel, had set off an atmospheric nuclear explosion in the South Atlantic near Antarctica. The Air Force didn't believe the DIA report and asked for a second opinion. Our committee included one of my most valued young associates, Richard Muller.

The records on which the DIA report was based came from an

orbiting satellite with two light sensors programmed to look for the typical double flash of a developing nuclear fireball. Previously such light signatures always had been confirmed by the finding of fission products in the atmosphere. No radioactivity had been found to back up the new satellite data.

The two ways to read the one good satellite record from which the explosive yield was determined didn't agree as well as usual. Drawing on my bubble chamber experience, I asked to see a selection of the satellites' "zoo-ons," events so strange they belonged in a zoo. This idea was new to the DIA, but since their records were stored on computer tape they needed only a week to put their zoo together. Rich, Dick, and I found a steady degradation in record quality among these zoo-ons from confirmed explosions to events at which no one would look twice. Although the event we were studying had some of the characteristics of a nuclear explosion, only one of the two satellite sensors recorded it. Moreover, there was no indication from earlier or later records that the sensor that failed to record the event was malfunctioning. Both sensors looked at a large area of the earth's surface, so it was hard to believe that one sensor could see a nuclear blast and the other could not.

Someone on the committee proposed that a micrometeorite might have struck the satellite and dislodged a piece of its skin. Reflecting sunlight into the optical system of one sensor but not into that of its neighbor, the debris might have caused the questionable event. We constructed a believable scenario based on the known frequency of such micrometeorite impacts that reproduced the observed light intensity and pattern.

I doubt that any responsible person now believes that a nuclear explosion occurred because no one has broken security, among South Africans or elsewhere. U.S. experience teaches that secrets of such import can't be kept long. After the United States tested its first megaton-scale thermonuclear weapon, which completely evaporated the small Pacific island of Elugelab, stories about a disappearing island reached U.S. newspapers as soon as the task force steamed into Pearl Harbor and sailors had time to call home.

Many people think that solving a scientific puzzle is an exercise in logic that could be carried out equally well by a computer. To the contrary, a scientific detective's main stock-in-trade is his ability to decide which evidence to ignore. In our DIA briefing we were shown, and quickly discarded, confirming evidence from a wild assemblage of sensors: radioactive Australian sheep thyroids, radiotelescopic ionospheric wind analyses, recordings from the Navy's sonic submarine-detection arrays that supposedly precisely located the blast from patterns of sounds reflected from bays and promontories on the coast of Antarctica.

Inevitably, then, I had a real sense of déjà vu when the House subcommittee's acoustics experts pinpointed the location of the open mike by triangulating the reflections of supposed gunshot sounds from the buildings in Dealey Plaza. The National Academy of Sciences committee showed conclusively that the motorcycle with the open mike wasn't even in Dealey Plaza at the time the tape was recorded.

Most of us on the academy committee were astonished when we finished our work that not one of the six experts involved ever admitted an error or challenged our published findings. Nor was the House subcommittee reconvened to reconsider its conclusions. No more than the acoustics experts did the House members want to admit that they were mistaken. The prime-time television news coverage our report received nevertheless convinced many who had previously doubted the Warren Report.

FIFTEEN

Impacts and Extinctions

THE MOST recent development of my scientific career has also been the most unusual. I will probably be remembered longest for work done with my son Walt in a field about which I knew absolutely nothing until I was sixty-six years old. The field is geology; the work is our impact theory of mass extinctions.

Walt earned his Ph.D. in geology at Princeton. That required fieldwork, which he carried out on Colombia's remote Guajira Peninsula. After graduation he worked in The Hague, Netherlands, and then for the prospecting arm of a multinational oil company in Libya until Qaddafi expelled all Americans. In order that he and his wife, Millie, could escape the golden handcuffs with which industry sometimes shackles its best scientists, I had advised them not to raise their living standards to match Walt's salary. They took my advice, so when Maurice Ewing invited Walt to join the Columbia University Lamont-Doherty Geological Observatory at a small fraction of his oil company salary, he was in a position to accept. My son and daughter-in-law enjoyed academic life. When Doc Ewing died, however, their future at Lamont-Doherty became unsure. Around that time I happened to be talking to a Berkeley geology professor at a social gathering. He told me he admired Walt's work in paleomagnetic and structural geology. He'd like to have Walt in his department, he said, but he could offer him only an assistant professorship and thought he'd certainly refuse. I suggested he talk to Walt. He did. Walt accepted and four years later became a full professor.

Walt's decision rejuvenated my scientific career. Over the years we'd developed the ability to talk in detail about our scientific work, an experience I'd had with few nonphysicists. I tried to avoid such technical talk because I assumed it would bore outsiders. The close personal relationship Walt and I enjoyed dissolved the cross-disciplinary barriers. He told me about his paleomagnetic studies in the Apennines of central Italy, and I read the papers he and William Lowrie wrote. I didn't understand geologists' jargon, but I learned their techniques for measuring the residual magnetism of rocks and what those measurements revealed about the earth's changing magnetic field.

One day Walt presented me with a complicated rock the size of a cigarette package that was preserved in transparent plastic. He had chiseled it from the wall of a gorge at Gubbio in the Apennines of Italy. Its lower section was white limestone; its middle, a half-inch layer of clay; its upper, red limestone. Walt said the clay was laid down on the ocean floor sixty-five million years ago at the boundary between the Cretaceous and Tertiary periods—the K/T boundary—when the dinosaurs and most other forms of life on earth abruptly became extinct. With a hand lens he showed me in the white limestone the profusion of tiny shells of foraminifera, shell-forming protozoans that live in the sea. Above the clay layer in the red limestone, none were visible. The clay layer was seen worldwide, Walt added, as was the abrupt foraminifera disappearance.

That was one of the most fascinating revelations I'd ever heard. I asked why the clay layer had formed and how long its formation had taken. The limestone, Walt said, was 95 percent calcium carbonate from compacted seashells and 5 percent clay from continental erosion. He guessed that the calcium carbonate had stopped settling, leaving only the clay for the clay layer, and then restarted. No one knew why. On the basis of his scenario and the limestone sedimentation rates, Walt concluded that the clay layer probably took about five thousand years to form. (As we'll see, that was wrong.)

I had not found Walt's field of science exciting before; now I did. I suggested we devise a way to measure directly how long the clay layer took to form. We first thought of looking in the clay for measurable traces of radioactive isotopes created in the atmosphere by cosmic rays, but only beryllium 10 came close, and its half-life of one and a half million years was ten times too short.

Then I remembered that meteorites are ten thousand times richer in elements from the platinum group than the earth's crust (because the earth's radioactivity melted the early crust and the iron that fell down to form the core scrubbed out these siderophilic—iron-loving—elements by an alloying process). On a visit to Antarctica I had learned that one can detect meteorite debris in the ice there by measuring siderophile levels. I suggested we look similarly for extraterrestrial platinum in the clay layer. A quick check showed that iridium, atomic number 77, was a better choice; it has an enormous cross section for neutron capture, which suits it better to trace-element analysis. (Later we learned that two men at the University of Chicago had independently invented this method of measuring sedimentation rates. It didn't live up to their expectations or to mine. Fortunately I hadn't heard of their work; if I had, I'm sure we wouldn't have bothered to look for iridium at the K/T boundary.)

The Berkeley Radiation Laboratory is a wonderful place because of, among other reasons, the diverse skills and disciplines of its expert staff. As soon as Walt and I decided to measure sedimentation rates across the K/T boundary, we enlisted Frank Asaro and Helen Michel, nuclear chemists, as coworkers. Frank and Helen had set up one of the world's finest facilities for neutron activation analysis—bombarding a sample with neutrons to make some stable element (iridium, in our case) radioactive so that one can calculate its abundance by measuring the radiation level. They had recently proved with neutron activation analysis that the university's treasured Plate of Brass, which had been discovered many years before in Marin County and which was believed to have been posted by Sir Francis Drake to claim Western America for Queen Elizabeth I, was fake.

The Columbia University metallurgist Colin Fink had authenticated the Plate of Brass with chemical tests. In the mid-1970s its authenticity came into doubt. That it was exactly 0.250 inches thick and appeared to have been rolled rather than pounded troubled Ed McMillan. Frank and Helen drilled a few small holes in its back, exposed the cuttings to slow neutrons, and found antimony and arsenic levels a thousand times less than the best chemists of Drake's day could have achieved. So the Plate of Brass joined Piltdown Man as another historic scientific fraud.

Frank and Helen analyzed our sample. They showed, to our very great surprise, that three hundred times as much iridium was concentrated in the clay layer as in the limestone layers above and below. They showed further that the gross chemical composition of the clay in both limestones was identical but quite different from that of the layer clay.

We then had a mystery: the origin of the clay layer iridium appeared to be extraterrestrial. We first guessed that it had come from a nearby supernova explosion. Some astrophysicists had proposed, and some paleontologists had accepted the proposal, that the K/T extinction had been triggered by such an explosion. Since elements heavier than nickel are believed to be made in the neutron-intense environment of stellar explosions, other heavy elements should have been present in anomalous amounts in the boundary clay. The key isotope to look for would be plutonium 244, with a half-life of 75 million years, which had decayed away in ordinary rocks formed 4.7 billion years ago with the solar system but should still be present in the clay layer if it contained the residue of a supernova explosion, in which Pu-244 would have been freshly formed.

Supernova nucleosynthesis calculations predicted the Pu-244-to-iridium ratio that Frank and Helen quickly detected. We were terribly excited because the activated Pu-244 signature was unambiguous; several X-ray and gamma-ray lines had the correct energies and correct relative intensities. I telephoned the president of the National Academy of Sciences and asked him if I could describe our discovery at the next annual meeting, which would celebrate the U.S. Geological Survey's hundredth anniversary. He put me on the program and congratulated us.

Frank and Helen are careful scientists. Despite their certainty that they had detected the anticipated amount of Pu-244 in the boundary clay, they repeated their entire difficult and tedious procedure with a second sample of the same clay to confirm their results. To their great consternation they found no Pu-244. Searching for an explanation of this remarkable difference, they remembered that before their first assay a sample of manmade Pu-244 had been bombarded in the Radiation Laboratory's heavy-ion linear accelerator. They realized that a small Pu-244 particle could have been dislodged from the target, drifted in through their laboratory window, and come to rest in their tiny beaker. Penicillin, after all, had been discovered in just that way.

I immediately canceled my talk, of course, but I was faced then with the problem of what to say about our experiments. If the newspapers heard that the laboratory allowed measurable quantities of plutonium to float through the countryside, the Berkeley city council might well petition to shut us down. It wouldn't have mattered that the speck of plutonium emitted only one alpha particle a day and couldn't have been detected by the most sensitive radioactivity monitor ever built. In the article we were preparing, published in *Science* in June 1980, we showed both spectra and said that the first sample had been spiked with Pu-244; our neglecting to mention how the spiking had been accomplished took nothing away from the statement. Frank in any case realized later that the accidental spiking had to have happened inside a well-protected radioactive-materials-handling hood, where the plutonium had every right to be. No one had been careless.

In the spring of 1979 Walt gave two papers on the large and unexplained iridium enhancement and showed with our plutonium data that the iridium didn't come from a supernova. He then returned to Italy to collect more boundary samples. I worked on theory, trying to identify a suitable source for the iridium and a mechanism for the extinctions. We felt we had demonstrated that the iridium was really extraterrestrial in origin. Not everyone immediately agreed.

Chris McKee, a Berkeley colleague, served through the summer as my astronomy consultant. Chris suggested the iridium could have been deposited by a ten-kilometer asteroid hitting the earth, but neither he nor I could invent a believable killing scenario. An object of that size smashing into the ocean would create a giant tidal wave, but such a wave couldn't kill dinosaurs in Mongolia and Montana. (Our *Science* paper didn't carry the acknowledgment we originally wrote of Chris's contribution, because a referee said the idea was obvious.)

I set to work to study the astronomy literature to find a killing mechanism. I speculated that the solar system might have passed through a giant cloud of molecular hydrogen, leading to the combustion of the incoming hydrogen by atmospheric oxygen and killing the animals by anoxia. Such a passage would take longer than the geological record allowed, however. I looked into the possibility that the sun had become an ordinary nova—the energy release would have explained almost everything—but then I learned that only a member of a pair of close double stars can become a nova. I even attempted to highjack the hydrogen and iridium from Jupiter, but I couldn't make that scenario work either. It wasn't at all obvious what brought in the pulse of iridium and killed most of the creatures on earth. Frank Asaro has recalled that I invented a new scheme every week for six weeks and shot them down one by one.

Finally I took a fresh look at incoming asteroids and comets. I first considered such an object passing through the atmosphere but missing the earth's surface, thinking it could fragment into dust that would be captured by the earth's gravity and eventually fall onto its surface. For a model I used the comet that fragmented over Tunguska, Siberia, in 1908 and destroyed many square miles of forest. But I could easily show that the deceleration of a ten-kilometer diameter asteroid or comet passing through the atmosphere was the same as the acceleration of gravity on the earth—not sufficient force to cause fragmentation. Because the Tunguska comet was much smaller, its deceleration was correspondingly greater, so it burned up.

But I was at last on the right track. A large body could have struck the earth with enough energy to vaporize itself and many times its weight of rock. Atmospheric dust would condense from the vaporized rock and fall out like the dust from the 1883 explosion of the Krakatoa volcano.

My father had given me the large 1888 Royal Society volume on Krakatoa. I had passed it on to Walt. I now borrowed it back and studied it. It told everything known then about the explosion, including how many cubic miles of volcanic rock had been blasted into the stratosphere and how it drifted back to earth. Sir George Stokes, whose viscosity law enables us to calculate how rapidly small particles fall in the air, measured the size of the Krakatoa dust particles by observing the diffraction halos they made around the moon. He calculated that the dust should remain aloft for two years, the length of time the dusty sunsets persisted all over the earth.

I concluded that a ten-kilometer piece of solar system debris hit the earth sixty-five million years ago and threw dust into the stratosphere that made the sky dark as midnight for several years, thereby stopping photosynthesis and so starving the animals to death. (We calculated the diameter of the object easily from the amount of iridium observed in our small sample plus the tabulated concentration of iridium in meteorites; in the past six years no one has suggested a better value than our original ten kilometers.) In the K/T extinction all land animals weighing more than fifty pounds disappeared from the fossil record; the environmental crisis I proposed would certainly have been harder on the larger and therefore less numerous animals. I thought that it would also get cold during this prolonged midnight, but I wasn't sure. If the sun suddenly stopped shining, it would doubtless get cold, but in the scenario I was considering the normal amount of sunlight would still hit the top of the atmosphere. The dust grains would absorb more sunlight than the atmosphere normally does and might actually heat it up. Extensive computer modeling has since shown that the dust would drop the temperature to zero degrees Fahrenheit for six to nine months. In effect, all the animals were transported to Antarctica for that time.

Frank, Helen, and I phoned Walt in Italy and told him that our impact scenario explained everything known about the K/T extinction. Walt also found it convincing and sounded excited. However, when Frank and I proposed attending a paleontological meeting in Copenhagen to which Walt had been invited to discuss the K/T extinction, he turned surprisingly negative. We thought the paleontologists would be delighted to learn what caused the extinctions. Walt knew better and correctly urged us to stay home.

We started instead on a long paper. By the end of the summer, when Walt returned to Berkeley, we had gone through many drafts. Walt did important rewriting so we wouldn't sound amateurish to our geologist readers, and we circulated a preprint among geologists and paleontologists in November. Several geologists repeated and confirmed our iridium measurements at different K/T boundary locations. Malvin Ruderman, a proponent of the supernova theory, responded immediately to our preprint with a brief and pointed note. "Dear Luie," he wrote: "You are right and we were wrong. Congratulations." Mal's response exemplifies science at its best, a physicist reacting instantly to evidence that destroys a theory in which he previously believed.

When we sent the paper to *Science,* Phil Abelson, the editor, returned it. It was too long, he said, and *Science* had published several papers in recent years purporting to explain the K/T extinction. "At least n-1 of them must be wrong," Phil remarked. He finally agreed to publish a shortened version of our paper, a courageous decision that earned our gratitude.

Our impact theory of extinctions is a fine example of team science. We needed Walt's geological expertise, Frank's and Helen's nuclear and chemical competence, and my background in physics and astronomy. If any of this knowledge had been unavailable, the theory might have been a long time coming. Discoveries are often made simultaneously by two or more groups because the time is ripe for them. I don't believe this was the case with the impact hypothesis.

I gave the first talk on the impact hypothesis at the American Association for the Advancement of Science annual meeting in January 1980. Science writers reported the story throughout the world, and most of the feedback was favorable. Some paleontologists strongly opposed the theory at first; a few continue to do so. Earth scientists liked it because it made many predictions that could be tested. Many laymen apparently believe that scientific theories can be verified experimentally. That's not true. Theories make predictions that can be confirmed, which leads to a theory's acceptance. But most theories can never be proved, only disproved. A classic example is Newton's theory of gravitation, which withstood detailed scrutiny for more than two centuries but eventually flunked three famous tests that Einstein's theory passed. One of the few theories that was certainly confirmed was proposed by Copernicus: the sun doesn't orbit the earth once each day; that apparent motion is caused by the rotation of the earth on its axis, as a high-quality gyroscope can easily demonstrate.

Fifteen of our surprising predictions were confirmed. In a few cases where the evidence was different from our expectation, our theory could be adjusted easily and actually strengthened.

We predicted that the boundary clay was different from the clay in the adjacent limestones above and below, which we said would be identical. We predicted that the iridium enhancement would be seen worldwide, as were the boundary clays and extinctions. By 1986, iridium enhancements had been found at nearly one hundred sites by nine groups, and only at two or three paleontologically defined K/T boundaries was there no iridium, a remarkable batting average in a field plagued by hiatuses—missing sections of the geological record. The relative abundance of twelve elements at the various locations seldom differ by more than 30 percent, whereas the element-to-element differences are as much as a factor of one million. One of Frank's and Helen's most impressive observations was that boundary clay from a deep-sea drilling core in the Pacific agreed in composition with a Danish boundary clay just as well as two Danish boundary clays from sites a mile apart agreed, while other local clays at both widely separated locations differed greatly from the boundary clay the two sites had in common.

Our theory predicted that the relative abundances of siderophilic elements in the boundary clay would be the same as those for solar system debris, the most common of which is stony meteorites. This was confirmed by R. Ganapathy. Frank and Helen found abundance ratios for iridium, platinum, and gold that were consistent with those of meteorites but quite unlike those of any crustal rocks listed in the literature.

Some geophysicists argued that the boundary clay iridium could have been precipitated by a sudden change in ocean chemistry rather than brought in by an asteroid. We countered with a prediction that iridium would be found in rocks formed on continental sites, far from any ocean. We prepared a proposal for a federal grant to examine a Montana site. The proposal was rejected because a peer reviewer reported, in effect, that we'd be wasting money looking at a continental site since everyone knew the iridium was precipitated out of the ocean chemically. We looked in Montana anyway, but just before we found iridium there, Carl Orth and his group at the Los Alamos National Laboratory identified it in a drill core they pulled from New Mexico's Raton Basin. They subjected their drill core sample to neutron activation analysis, and the iridium suddenly increased by a factor of three hundred precisely where the paleontologists told them to look. That exciting measurement proved that the K/T iridium didn't come from the ocean. It was deposited on the continents as well as on the ocean floor. The Los Alamos discovery added great strength to our theory.

A ten-kilometer asteroid hitting the earth would release an energy of some one hundred million megatons of TNT equivalent. A one-megaton hydrogen bomb is a big bomb. The worst nuclear scenario yet proposed considers all fifty thousand nuclear warheads in U.S. and Russian hands going off more or less at once. That would be a disaster four orders of magnitude less violent than the K/T asteroid impact. It was quite simply the greatest catastrophe in the history of the earth of which we have any record, and by now the record is well documented.

That understanding led us to suggest that plants as well as animals must have suffered from the impact's environmental shock. Leo Hickey, a Washington paleobotanist and a very good friend—Walt and he went to graduate school together—hotly disputed our prediction. He said he found no evidence that plants had noticed the impact. Carl Orth's team showed to the contrary that precisely at the level in the Raton Basin rock where the iridium concentration increased

300-fold, the fossil pollen density fell to 1/300 its usual value. The two different spikes occurred at the same level to within a centimeter. Stratigraphic measurements have rarely shown such high spatial resolution. The narrowness of the pollen dip explained why paleobotanists had not seen it; they collected their samples farther apart and in the process simply jumped over it.

We predicted that the iridium enhancement would appear just above the highest (and last known) dinosaur fossils. The dinosaurs were reptiles; those on land at the end of the Cretaceous period dramatically and catastrophically disappeared. Several orders of reptiles, including giant marine reptiles, went completely extinct. Normally extinctions take place at the species level. The passenger pigeon disappeared at the end of the nineteenth century. The condors will probably go extinct soon. Each loss is a species extinction. The next level up is genus; above that, family; above that, order. For animals the only higher taxa are class and phylum. A catastrophe that suddenly wipes out several orders is a spectacular event, not to be attributed to some ordinary environmental change, as some of my paleontologist colleagues believe. Dinosaurs were the largest animals that ever lived on land. There were large reptiles in the sea, the plesiosaurs. There were large reptiles flying around in the air, the pterosaurs. All disappeared suddenly, never to be seen again.

Bill Clemens, a Berkeley paleontology professor, collected samples for us from a Montana site where he excavated dinosaur fossils. Frank and Helen found the iridium layer three meters above the most recent dinosaur bone Bill had turned up. Bill has insisted that this three-meter gap (later reduced to two meters) shows that the dinosaurs went extinct 20,000 years before the asteroid impact. We are sure the gap is a sampling error, because the average vertical spacing between finds of dinosaur fossils, which are rare, is one meter. We know the marine fauna disappeared precisely when the boundary clay was deposited, because their remains are so plentiful that their average vertical spacing is less than a millimeter. I am unable to take seriously Bill's notion that the dinosaurs, after ruling the world for 140 million years, suddenly and for no particular reason disappeared just 20,000 years before the greatest catastrophe ever known to have been visited upon the earth. Sampling errors do occur, and the chance of their occurrence is much higher than the ratio of 20,000 to 140 million years. (This ratio corresponds to betting odds of

7,000 to 1, whereas the odds against the sampling errors are close to even money.) The evidence is strongly against Bill Clemens's theory and strongly in favor of ours.

I proposed to look for fossils above and below the iridium layer by means of high-frequency sonar, which is rather like Superman's X-ray eyes. That would greatly increase the number of detected fossils and decrease their average vertical spacing. We were sure that such an exploration would make the two-meter gap disappear. But there were so few people left who would change their positions because of any results we might obtain that we didn't develop this interesting technique. We held long and interesting meetings with Bill and several other paleontologists every Tuesday for three months, mutually profitable encounters without personal conflict, but in the end we tacitly agreed to disagree. Bill finally accepted the impact theory but denied its connection with the sudden dinosaur extinction, for which he offers no competing explanation. (I once said, "Bill's theory is that our theory is wrong.")

We predicted that the boundary clay should show signs of exceedingly high temperatures and pressures from the impact shock (these were actually postdictions; they followed directly from the theory, but we hadn't the temerity to articulate them before other groups observed them). These signs would be absent if the iridium came from volcanoes. Sand-sized sanidine spherules turned up in the boundary clay, high-temperature creations similar to microtektites. A rare form of crystalline quartz seen only in rocks that have undergone shock metamorphosis demonstrated the effect of high pressure: it had been found previously only near impact craters and underground nuclear explosions. The shocked quartz convinced the few remaining geological skeptics; it was evidence they understood and found persuasive.

Our harshest critic was Bob Jastrow, a Berkeley-trained particle theorist who passed through astrophysics and space science to become a geology professor at Dartmouth College. In a long article in the magazine *Science Digest,* Bob ripped our work to shreds. When I read it, I called him. We had a long, surprisingly friendly talk. I pointed out what I saw as his many errors of fact. Afterward I started a letter to the editor refuting Bob's criticisms, but Rich Muller made me realize I couldn't make a convincing case in the space available and that my letter would give Bob the last word without fear of rebuttal. No second round is possible in such a forum. *Science Digest* eventually published a long and exceedingly favorable report on our work and its increasingly wide acceptance in the scientific world.

One part of our theory had to be modified, but the modification made it fit the paleontological record better than it had. I thought that the asteroid or comet (we don't yet know which it was) would vaporize itself and ten to a hundred times its weight of rock. The resulting fine dust would fall out over a period of three years, which would give the winds of the stratosphere time to transport it worldwide. Brian Toon of the NASA Ames Laboratory used a computer simulation to show that the material would fall out much faster—in a few months. The reason was that the dust particles would collide and clump together, and larger particles, according to Stokes's law, fall faster than smaller ones—too fast, it appeared, for worldwide transport. We might have faced serious trouble with Brian's new data except for the comforting fact that we had already seen the iridium layer worldwide and knew there had to be a transport mechanism.

We were wrong about stratospheric transport because we accepted Stokes's Krakatoa dust observations at face value. Unknown to him, the Krakatoa dust fell out in three to six months, but the job of obscuring sunsets was smoothly taken over along the way by much finer aerosols that accompany volcanic eruptions. No such aerosols accompany impact explosions.

Material suspended in the atmosphere needs more than a year to spread from the Northern Hemisphere to the Southern, on the evidence of the Soviet hydrogen bomb tests of the 1950s, which created large quantities of carbon 14. We required some more rapid system of transport.

We pointed out our problem to the crowd at the 1981 Conference on Large Body Impacts and Terrestrial Evolution at Snowbird, Utah, where Brian Toon had raised the question. The next day two groups reported their calculations that showed the material was spread not by stratospheric winds but by either of two much faster mechanisms. A Los Alamos group demonstrated that it actually went into ballistic orbit, like a crowd of ICBM's, and spread worldwide in an hour. We had known that the impacting body brought in enough energy to orbit its debris, but couldn't see how to move all those little particles up through the atmosphere. The Los Alamos group, and also one at Pasadena, used large computers to simulate the impact event. It turned out that convective vertical winds in the fireball could do the job. Both groups showed that when the asteroid hit it would distribute the material worldwide very rapidly and that it would be diluted by between twenty and a hundred times its incoming weight. (The measurements point to a dilution factor of about ten.)

The computers very nicely got us out of trouble. It happened that the Pasadena group was actually wired in by a special line to the Berkeley computer. Down in the basement of our building it was cranking away on our problem, and we didn't even know it.

Marine paleontologists had been troubled by our proposed three years of darkness because so long a period would kill off most plant and animal life in the sea. They were relieved to look at only three to six months of darkness instead. Hans Thierstein, a paleontologist who specializes in surface microplankton that photosynthesize, told the Snowbird conference that the shorter period of darkness provided a good mechanism to account for the observed fossil pattern. Good theories survive and are strengthened by glitches such as our darkness problem; bad theories go bankrupt.

Another of our predictions has had its ups and downs. Eugene Shoemaker, an expert on earth-crossing asteroids and moon craters, calculated that a ten-kilometer asteroid should hit the earth once every 100 million years. There have been five major extinctions since the Cambrian period began, about 500 million years ago. We suggested that all these extinctions could have been triggered by asteroid or comet impacts. Frank Asaro discovered an iridium layer at a minor extinction, the Eocene-Oligocene boundary, 35 million years ago. The E/O layer appears worldwide in the earth's equatorial zone and coincides not only with the disappearance of some radiolaria (foramlike creatures with chemistry based on silicon rather than calcium) but also with a layer of microtektites. Tektites, smooth glassy objects that range in size from centimeters down to fractions of a millimeter, are generally thought to result from meteorite bombardment. Our E/O discovery tied iridium directly to extinctions and to impacts.

We tried hard to find iridium at the 240-million-year Permian-Triassic boundary. The best samples we had came from China, two sets from different locations near Nanking. After a year we had not yet succeeded, although Frank and Helen had pushed their detection limit down to a few parts per trillion. The gross chemistry of the clay layer at the putative P/T boundary was different from the chemistry of the rocks above and below it, but there was no detectable iridium. Walt thought the samples might have been taken from the wrong place, and we have joined forces with a Chinese group to search for iridium in a wide band centered on the layers where we and they have previously looked. In the meantime Carl Orth at Los Alamos found an iridium enhancement at the Late Devonian (F/F) extinction, another of the five majors, but he thinks it was unrelated to an impact, and of biogenic origin.

How widely is the impact theory accepted? Science has no supreme court to decide such matters, but most American scientists would defer to the National Academy of Sciences. The academy sponsored a television series in 1986, the first time to my knowledge it had ever done such a thing. The fourth segment of the six-part "Planet Earth" series treated extinctions and cited the impact of an asteroid or comet as the cause of the K/T extinction. It referred to no other theories, and the only scientists interviewed in that segment of the program were Walt, Gene Shoemaker, and I.

Some earth scientists complained that we have looked for iridium only where there are extinctions and that we might otherwise find it elsewhere, perhaps everywhere, in the geological record. Frank and Helen have searched through long stretches of sediments and have found iridium at the level of only thirty parts per trillion elsewhere than at extinction boundaries, the amount expected from the steady infall of meteoric dust. Those measurements have been time consuming, taking typically a full weekend per sample. But in mid-1986 the search program underwent a sharp speedup. We began to be able to make a measurement in seven minutes and would soon begin a systematic search through all of geological time at a sensitivity of fifty parts per trillion of iridium. We hoped then to examine some thirty thousand samples per year, and Walt has organized collection teams to keep up with the voracious appetite of the new detection equipment, which has an automatic sample changer and will work around the clock.

My extensive knowledge of nuclear physics led to the invention of the new technology; Frank's expertise in instrument design brought the new device into being. One prominent neutron activation analyst who reviewed our proposal for funding the new instrument wrote that "unless the Berkeley group shares its analyzer with others in the iridium detection business, it will drive everyone else out of the field." As followers of the Ernest Lawrence tradition, we had already written to others in the field offering them the use of our new detector.

The latest scientists to challenge the impact theory were Charles

Officer and Charles Drake of Dartmouth College. They published two articles in *Science* claiming that volcanic activity explained the experimental evidence better than an impact. We answered all their points in *Science,* but in late 1985 Officer gave a paper at the San Francisco meeting of the American Geophysical Union saying the same thing. One of his major points was that although Walt's student Sandro Montanari had found sanidine spherules in Italy's Gubbio region only in the clay boundary layer, he and Drake had found them almost everywhere above and below the clay layer as well, in sharp disagreement with our team's observations. He showed a slide of the spherules, which were smooth and hollow.

In the two-minute rebuttal that Walt was allowed, he said that the discrepancy between the work of our two teams could be explained by the fact that Officer and Drake hadn't cleaned their samples of surface contaminants. Sandro had seen their widespread "sanidine spherules" and demonstrated that they were modern insect eggs—even giving their Latin species name. He showed that these spherules burned in air, didn't dissolve in acids, and could be dented if poked with a pin.

At that point the audience of several hundred earth scientists burst into laughter, something I'd never witnessed before in my fifty-three years of attending scientific meetings.

Late in 1983 two Chicago paleontologists, Dave Raup and John Sepkoski, offered evidence that during the past 240 million years extinctions have occurred about every 26 million years. I hoped they were right but found their evidence unconvincing. I did so partly because I was unable to think of a mechanism to explain the apparent periodicity. (This is an example showing me doing bad science; I should have concentrated on the evidence.) I thought that the impacts must be random because the number of earth-crossing asteroids that Gene Shoemaker saw from Mount Palomar (and that had to be random) agreed with the cratering rates on the earth and the moon.

I showed Rich Muller, my former student, a draft of a letter I had written to the Chicago paleontologists showing why they had to be mistaken. I asked him to play devil's advocate and identify any errors I had made. At first he agreed with me, but later he came to believe the periodicity was valid and set out to find an explanation. After trying half a dozen astronomical explanations that he eventually concluded were unsatisfactory, he proposed that the sun, like most stars in our galaxy, has a companion star and that the two orbit their mutual center of gravity every 26 million years. In the final stages of this work, Rich had advice and assistance from Marc Davis of the Berkeley physics and astronomy departments and of Piet Hut of the Institute for Advanced Study in Princeton. They named the companion star Nemesis; they proposed that it passed periodically through the Oort comet cloud, which supplies new comets to the inner solar system to make up for the old ones that are ejected by gravitational encounters with the giant planets Jupiter and Saturn. Nemesis' passage would perturb some of the Oort comet orbits gravitationally enough to send a pulse of a billion fresh comets into the inner solar system, where perhaps twenty-five would by chance hit the earth.

When Walt heard Rich's idea he realized that earth craters should also show a 26-million-year periodicity. From a tabulation of the sizes and ages of large known earth craters, he plotted those with the smallest dating uncertainties as a function of time. He told me in great excitement that the periodicity jumped out at him. He and Rich teamed up to conduct a more sophisticated computer analysis of the cratering data and informed me that their best period was 28.5 million years. Playing the devil's advocate in these deliberations, I immediately pointed out that the difference between 26 and 28.5 required a phase shift of one full cycle in ten. "Everybody will laugh at you," I challenged them. They reexamined everyone's data carefully and concluded that the Chicago period was consistent with 28.5 million years if two minor extinctions were combined, which the dating errors accommodated. I was delighted to watch two of my closest friends working together for the first time, and so productively.

Some of the ideas described in their two Berkeley preprints had been developed independently by three other groups. As a result, *Nature* published all five papers together in the same issue. At Dave Raup's suggestion he and I cohosted a small workshop on periodic extinctions in Berkeley. Almost everyone active in the field attended. Gene Shoemaker spent an entire afternoon telling us why no one should believe in "Rich's star." He and Walt had dinner at a Chinese restaurant that night; when the fortune cookies arrived, Walt's contained the message "The star of riches is shining on you."

Whether it was or not remains to be seen. While Rich worked on his papers, I asked if I could start a search for Nemesis. He was pleased to have me do so. I contacted Dave Cudaback of the Berkeley astronomy department. Together with Saul Perlmutter, one of Rich's graduate students, we began searching star catalogs. I learned a great deal of astronomy in the next few weeks. When I brought Dave what I thought was a startling plot of luminosity relationships among the stars nearest the sun, he informed me that I had just rediscovered the main sequence, a major grouping of stars.

On the basis of a 28.5-million-year periodicity, we knew that Nemesis was no more than 2.5 light-years away, making it the closest star to the earth. It would therefore move back and forth against the background stars by three arc-seconds of parallax as the earth went around the sun. In order not to be obvious at such a small distance, it had to be a dim red or even a brown dwarf. Saul searched through two star catalogs by computer and found no candidates. His computer refused to decode a third catalog tape; I searched all nine thousand entries in the printed catalog by hand and eliminated every one.

It occurred to Rich Muller that his automated telescope program that searched for supernovae in their earliest stages was ideally suited to search for Nemesis as well, looking among dim red stars to find the one with the largest parallax. Dave Cudaback and I continue as guest scientists in Rich's group; we hope to be useful members of the team that eventually finds this great scientific prize. That success is of course not assured. Only a small fraction of knowledgeable scientists think that Nemesis exists. Having looked closely at all the evidence, and in spite of my early skepticism, I now feel that the case for periodic extinctions and periodic impacts is strong. Of all the theories invented to explain these observations, that of the companion star seems to be the only viable one. But only time will tell the real story.

SIXTEEN

Introspection

I MADE my last landing and put my Cessna 310 up for sale in 1984, fifty years almost to the day after my first solo flight. When I stopped flying, I had current multi-engine and instrument ratings, frequently flying on instruments down through the overcast to land at airports I had never seen before. Unlike driving a car or playing the piano, during which I could think about other problems, flying demanded my full intellectual attention. I found few activities as satisfying as being pilot in command with responsibility for my passengers' lives.

I flew many military aircraft from the copilot's seat. After I completed the Navy's instrument course in 1942, I could fly any plane, regardless of size, using the controls to make the instruments do what I wanted. Once the pilot of a B–29 gave me the controls, and I put the plane into a well-coordinated 360-degree turn banked at 70 degrees, maintaining altitude to within forty feet. The pilot complimented me on my performance and asked casually if I knew the wings came off at 80 degrees.

I was the first civilian to make a low approach under the hood, on GCA. My check pilot, Bruce Griffin, would have taken over if I had done anything potentially dangerous. By the time I closed out my flying career, I had made hundreds of low approaches in my own plane on the modern ILS (instrument landing system), sometimes under the hood with a check pilot but frequently alone or with passengers under IFR (instrument flying rules) conditions, with no visual reference to the ground or to the horizon. Putting myself in hazardous situations is consistent with accepted behavior for an experimental physicist. Our unofficial sport, after all, is mountain climbing.

For some time after the war I didn't do any piloting. In 1959, on a commercial flight to Los Angeles, Jan and I found ourselves talking about planes. She said she thought learning to fly would be fun. When we landed we rented a car, but instead of driving to our hotel I drove to the Santa Monica airport and pulled up in front of a flying school. Jan wondered what we were doing; I said she was going to have her first flying lesson. With an instructor, a Cessna 172, and a husband in the backseat Jan took off and handled the controls flying up and down the coast over the ocean. After a year of sporadic flying lessons she learned to be a good pilot but never wanted to earn her license. Both my sons are pilots.

Most people don't know about the scrapes private pilots get themselves into, except when they kill themselves. Since this book is highlighting adventures, I'll describe one that's typical of those all of us with more than a thousand hours in light aircraft have experienced.

A few years ago my younger son, Don, and I were flying home from San Diego late at night. As we passed over Los Angeles, we couldn't see a single light through the thick undercast. As I scanned the engine instruments in the dimly lit cockpit, it appeared that the oil pressure on one engine had dropped to zero. Such a condition can quickly lead to an engine fire, so I immediately cut the ignition on that engine and feathered the propeller.

I didn't want to fly several hundred miles at night over California's coastal mountains on one engine. Don talked to the tower operators at Santa Monica, who said they could guide me to the proper point from which I could start on an ILS approach to their main runway. I've made many single-engine hooded approaches of that kind in good weather with an instructor checking me out for the semiannual renewal of my instrument rating. If I didn't perform perfectly under such conditions, the instructor could give me back my engine for a go-around and a second try. That wasn't possible with a dead engine, so I took the controller's suggestion that I divert to the John Wayne Airport at Irvine, which had no cloud cover. I could of course have handled the communications with the ground, but I was pleased to have seventeen-year-old Don there to help his seventy-one-year-old father.

As my instructor had taught me, I flew low, once over the runway, to check it for obstructions unknown to the tower and then made a smooth landing. I couldn't taxi to a tie-down position on one engine; a truck came out to tow us. Along the way I noticed that the dial that was reading zero wasn't the oil pressure gauge, as I had supposed, but another one of the three on that single instrument face. It indicated that the important cylinder head temperature had dropped to zero—an obvious impossibility. A circuit breaker had tripped. Although I had spent countless hours reading the airplane's instruction book, I was unaware of the existence of that particular circuit breaker. I quickly found it and reset it before the tow truck had finished its job, and the temperature gauge suddenly indicated a reasonable value. If that circuit breaker had ever tripped in the many hundreds of daylight hours I'd accumulated in Cessna 310's, I would immediately have known what to do over Santa Monica, and we would have flown on uneventfully and landed at Oakland two hours later.

As I recall this incident two years after I stopped flying, I'm struck by the hubris I had developed regarding my flying skills. All pilots feel that way, or they'd stop flying. We don't have to feel, as Chuck Yeager does—with good reason—that he's the best pilot who ever lived. But we do have to think we're pretty good. I really enjoyed my flying career, but my fiftieth anniversary was a good time to quit.

I'll finish this brief account of my involvement in aviation by mentioning a few of my aviation heroes. They made me welcome in my intrusion into their field. (If you feel I'm dropping names, then much of this book will have made you feel that way, because I've recalled my experiences with friends who are, or were, well-known and world-class physicists. Heroes have been important to my development as a scientist. I was disappointed never to have met my number one hero, Ernest Rutherford; as I've mentioned, he died in 1937 just before my work with tritium and helium 3 would have made him aware of my existence.)

In aviation my two principal heroes are Jimmy Doolittle and Chuck Yeager. Jimmy, with a doctorate in aeronautical engineering from MIT, was the first man to make a blind landing and was a daredevil racing and stunt pilot before his exploits in World War II made his name a household word. Chuck was our finest test pilot, the first person to break the sound barrier as well as a World War II fighter-pilot ace. As a result of Tom Wolfe's book *The Right Stuff* (and the movie made from it) and his own best-selling autobiography, *Yeager*, he's now a folk hero. I asked both Jimmy and Chuck to sign their biographies for me, the only autographs I've asked for in my adult life. Jimmy described himself as a "friend and admirer." Chuck called me "an old and wonderful friend."

Jan and I spent a total of several weeks at various times as house guests of Jackie Cochran at her beautiful ranch near Palm Springs. Jackie was the first woman to break the sound barrier (in the same two-seater F–104 in which I later flew); Chuck devotes a full chapter to her in his book. I've watched Neil Armstrong autograph piles of color photographs of himself on more than one occasion, and I could easily have asked for one for myself. I respect Neil's carefully guarded privacy; it never occurred to me to ask for such a gift.

My father thought of himself as a world traveler because he and my mother toured Europe once in 1931. In those days such a trip took two weeks by boat and was expensive. I've been a world traveler by more modern standards. I've walked on eight continents, visited 55 of the 157 United Nations countries, and seen 27 more. New Guinea is the largest landmass I've not yet walked on.

My brother and I arranged automobile trips in college to visit as many states as possible (I long ago visited my fiftieth state). Jan and I planned only one trip that way. In 1963 we traveled to Africa prepared to explore its length and breadth in one month. Formerly British, Belgian, French, and Portuguese colonies were moving through various stages of independence, and it was an exciting time. In Ghana, though, Jan was bitten by a mosquito, and ten days later, after we had visited several countries in southern Africa, she came down with malaria in Kenya. Her terrible chills and fever lasted ten days; we spent them in our hotel room.

I found I had ten days to fill without the bother of the telephone or the diversion of a functioning wife. My young postdoctoral colleagues had given us a new Canon zoom-lens movie camera to photograph the animals on our African tour. Keeping it steady at its longest focal length, forty millimeters, required bracing myself against a tree or a car. So while I attended Jan in Nairobi I set about designing an optical lens attachment that would eliminate the bothersome jitter and still allow the camera to be zoomed in focal length and its axis to be panned up and down or side to side. The civilian context of the challenge led me away from the straightforward but complicated and obviously expensive stabilized optics schemes I later learned the military used. I succeeded in my design effort and wrote a long disclosure to my patent attorney. I mailed it from Nairobi to establish the date of invention.

Our African trip resulted in the founding of a new industry, stabilized optics. Jan recovered quickly. Back in Berkeley, Pete Schwemin made a working model of the stabilized system I had designed, and we showed it to Chuck Percy, the president then of Bell & Howell. Chuck liked it and supported our Optical Research and Development Company (ORDCO) with a research and development contract. We produced half a dozen different stabilizing systems that Bell & Howell built into cameras and binoculars. Chuck wanted to market them, but he also wanted to run for the Senate. He succeeded, of course, and his successors, less venturesome, terminated our support. Eventually the patent I assigned to them expired. In 1981 Pete and I formed Schwem Technology. I am its chairman of the board and chief inventor. Jan is its full-time president.

I spent almost a year at a computer terminal learning to design highly corrected lens systems. That's a skill few physicists ever learn. I took it on as a challenge and found it interesting. The most important result of the exercise is that I can now communicate with our highly skilled optical consultant, Dick Altman. Dick does all our final lens designs. Schwem produces and sells stabilizing zoom lenses for the shoulder-mounted video cameras used in electronic news gathering. The large inertial helicopter mountings that serve for the production of television commercials are too expensive for local television stations. Our lens, the Schwem Gyrozoom, which works between 60 mm and 300 mm ("extreme telephoto"), is completely portable and priced to make it affordable for every TV news truck. It doesn't stabilize either the camera or the lens, but only the image being recorded. Dave Packard liked our demonstration videotapes and became a substantial Schwem investor.

I spoke with Dave's secretary, Margaret Paull, not long ago. She asked me what I was doing. "Working on dinosaurs," I replied, "and trying to make some money for Dave." "He doesn't *need* any more," she said, a conclusion based on having to clean his office of a torrent of begging mail after *Fortune* posted his picture on its cover as the U.S. citizen whose net worth had increased the most in the preceding ten months—about $1.2 billion. I've already told how Bill Hewlett supported our pyramid project. Now his partner's name turns up. That's very appropriate, since my introduction to business came in 1957 when Bill and Dave invited me to sit on the small board of directors of the small Hewlett-Packard company. I served in that capacity for a very interesting twenty-seven years as the company grew into a $4-billion-a-year giant. One year recently HP and IBM tied for first place in *Fortune*'s annual poll to determine which U.S. company American businessmen most admire. I reached the mandatory retirement age in 1984 (sometimes referred to as "statutory senility"), but at Bill's and Dave's request I continue to serve as a salaried consultant to HP Labs. Jan and I made many trips to HP plants in Europe and in the Far East, and we could have had no finer course in how to run a business than the one Bill and Dave gave us from a front-row seat. We have started two different optical companies in two quite different fields; without such fine tutors I wouldn't have dreamed of going into business. The first of the two companies was quite successful and was sold to a large pharmaceuticals company in 1979. Hewlett-Packard and its people deserve more mention here than they've had, which just shows that there are people in this world called editors.

The last paragraph speaks of two optical companies, but I've mentioned only the one in the field of stabilized optics. The earlier one exploited a very unusual type of lens I invented when I had to start wearing bifocals and decided that there should be a better way to handle my age-related inability to focus my eye lenses than the bifocals Ben Franklin invented two centuries earlier. I came up with a completely new kind of lens system in which the focusing was accomplished by moving two odd-shaped plastic elements at right angles to the line of sight. Although no one has yet marketed the system for its intended use in spectacles, our company, Humphrey Instruments, sold optometric testing instruments that used our "transverse optics," and the Polaroid Corporation introduced a new camera in 1986 in which the focusing is accomplished by such transverse lens motion.

We demonstrated the technology to Polaroid more than twenty years ago, and had looked forward to receiving royalties for its use, but the company avoided any royalty payments by waiting out the seventeen-year lifetime of the patent. It expired two years ago, so anyone is now free to use the invention. We sold Humphrey to Smith,

Kline several years ago—my first profit on any of my forty patents—and Jan, Pete, and I are back now as entrepreneurs, watching Schwem Technology move toward profitability. Recently I heard Bob Wilson, who directed Fermilab for many years, discuss the small company he's founded to sell superconducting accelerators to hospitals for therapeutic use. "I've spent my whole life in a socialistic society," Bob said, "getting money from the government to do things with no commercial value. So I find it exciting to enter the capitalistic world where your success is measured by the numbers on your financial statements." I'm happy that I didn't have to leave the world of science to dabble in the world of business. I agree with Bob that business is exciting and challenging, and as a hobby it certainly beats bridge.

Personalities often collide in the openly honest world of science. Hardly anyone is universally admired. Two exceptions come to mind from physics: Enrico Fermi and Willy Fowler. Enrico's awesome talents could have been frightening, but because he was gentle and friendly, people felt comfortable around him. Willy smiles easily and warmly and is truly interested in other people. Politicians display such characteristics; most physicists avoid acquiring them for fear of seeming insincere.

I am neither as smart as Enrico nor as outgoing as Willy, but I believe I'm widely reputed to be approachable and friendly. I base this assessment on two observations. First, when I'm introduced to young physicists from the East Coast or Europe they often say, "I'm glad to meet you, Luie." By no stretch of the imagination could I have said, "I'm glad to meet you, Bob," when Arthur Compton introduced me to Robert Millikan in my graduate-student days. Young physicists, technicians, and secretaries in the Radiation Laboratory all greet me by my first name, as I do them. Second, when the USSR and the People's Republic of China opened their laboratories to American physicists for the first time, in 1956 and 1973 respectively, I was a member of both U.S. delegations. Owen Chamberlain and I were the only Americans to be invited by both countries. I'm not universally loved. Some heartily dislike me. But all in all, I'm happy with my personal relationships.

I'm particularly pleased that the thirty-member Jason group continues to invite me to all its meetings as a senior adviser, the oldest of the four in that category. The Jasons are the brightest young physicists I've ever known, and they spend their summers working on highly classified problems involving national security. The need for such a group occurred to Charlie Townes and John Wheeler in the mid-1950s when they noticed that all the committees they served on relating to national security were populated with alumni from Los Alamos and the MIT Radiation Laboratory and getting older each year. They agreed that the younger generation should be introduced to such problems, and Jason was the result. I was forcefully reminded of the impact of Jason when I first served on an important committee chaired by Sid Drell, who was just too young to have worked in the laboratories where I was introduced to secret work and who "learned his trade" in Jason.

I was curious to watch Jan, as president of Schwem Technology, hire key people for our staff. The company advertised in Bay Area newspapers. Our personnel manager screened applicants, our operations manager interviewed the ten who seemed best suited to our needs, and Jan interviewed the finalists. Such thorough screening is necessary and standard, but at the laboratory I was able to operate differently. The last person I interviewed was Art Rosenfeld, who had just earned his Ph.D. under Fermi, in 1954. People have usually joined my projects by moving in, and getting to work, even when they have been committed to another project. They joined with me because they heard or knew from experience that I was fun to work with. This system was self-regulating, since they wouldn't move into my group unless they were confident of their ability to perform. Their freedom of movement also meant I've never had to fire anyone. If they didn't like the work or felt they had waded in beyond their depth, they could move out as easily as they had moved in. The creeping bureaucracy that has come to infect our government-operated laboratories now makes such utopian arrangements impossible.

In the near-anarchy of most laboratories where I have worked successfully, the manager had to earn the group's loyalty. Otherwise he woke up one morning to find his people gone. I'm proud that, like Ernest Lawrence, I generated loyalty in those who worked with me. Together we accomplished some outstanding work.

For many years, Ernest and I were the only physicists to have won Nobel Prizes for work done by large teams. More recently Burt Richter and Carlo Rubbia have joined us. If we four shared any common characteristic, it was an ability to lead groups into dangerous but rewarding new territory. There the leader's reputation was on the line, which helped motivate everyone on the team to make the project succeed. I've long felt a grudging admiration for Hernando Cortés. He was a despicable person, but by burning his ships to the waterline when he landed his small band of warriors in Mexico he showed them he was completely committed to its conquest. By contrast, I remember what happened to a friend who tried to start a new laboratory in the boondocks. He wanted to hire good people but had difficulty recruiting until he resigned from his university professorship. No one would fly on a commercial airliner if the crew showed up wearing parachutes.

I learned from Ernest to give lots of credit to my associates. I don't accept the Bible's theological teachings, but both Ernest and I followed its advice to cast our bread upon the waters. We both ended up with plenty of bread for ourselves.

If, as Carl Sindermann writes in *Winning the Games Scientists Play,* power is "a measure of a person's potential to get others to do what he or she wants them to do, as well as to avoid being forced by others to do what he or she doesn't want to do," then I have been powerful for some five decades. I'm perceived as a fighter for the things I believe in. Perhaps paradoxically, I see myself as someone who tries to avoid personal controversy. I feel I fight like the general who works from his command post behind the front lines. I can't think clearly in confrontations as a boxer, a policeman, or a frontline officer must. Even in less charged situations I usually let others do most of the talking. I'm articulate in front of groups who've demonstrated their interest in what I have to say by giving me the floor, but I've never learned to take it preemptively. I normally sit and listen. My social timidity surprises me, but that's the way I am. I make friends more often than enemies by being a good listener.

My name almost never appears on the countless petitions loaded with the signatures of Nobel laureates. I often find turning down petition requests difficult, but I promised myself on the morning of October 30, 1968, that I wouldn't become a "professional" laureate. I've seen no indication that such petitions from this side have lessened Andrei Sakharov's problems with his government. In fact, they may have exacerbated them.

Members of most professions can track their movement up the ladder of success relative to their friends and rivals by counting the honors they receive—Oscars and Emmys in the entertainment business and Pro Bowl invitations in the National Football League. Scientists are similarly aware of those who have received medals and awards and keep a mental scorecard to let them know how they are moving up in the esteem with which their peers view them. Military men wear their honors in full view on their uniforms, one ribbon for each. A naive observer counts the ribbons, but those of us who know the system recognize all the important ribbons and start from the upper left, since military honors, like military men, are ranked. When my former GCA test pilot, Pete Aurand, returned from duty in the Pacific in World War II, he wore only a single ribbon, not the three rows to which he was entitled. But the one he chose to wear was the prestigious Navy Cross.

The academic world emphasizes its honors by a similar inversion. Although an occasional scientist makes his *Who's Who* entry as long as he can, most of us in the know try to keep our entries as short as decently possible. Mine is sixteen lines long; half the list is honors—the Nobel Prize, the Collier Trophy in aviation, the National Medal of Science, the National Inventors Hall of Fame, and many more, including six honorary doctorates of science. Such honors are awarded by juries that poll possible nominators for appropriate candidates. One cannot apply; the statutes of the Nobel Foundation, in particular, say explicitly that all self-nominations will be ignored.

Valuing honors myself, I've worked hard to see to it that my favorite candidates win them as well. My first attempt and success involved Bill Brobeck, the designer of the Bevatron, on which I did my most important work. I was moved to action on the suggestion of one of Bill's social friends. I wrote a strong letter to Berkeley's chancellor proposing that Bill be awarded an honorary Sc.D. and then was disturbed to receive no reply. A year later I discussed this omission with a friend who knew the chairman of the honorary-degree committee. He volunteered to find out what went wrong. I learned that in the course of supporting Bill's candidacy his friend had emphasized Bill's extensive financial contributions to the university. Certainly Bill had supported the university financially, but emphasizing that fact raised a red flag in the degree committee; only less prestigious institutions reward financial angels with honorary degrees. When I learned what had gone wrong, I started the nomination process all over again. The committee started a new file and left the red flag behind in the old one. Sitting onstage at Charter Day ceremonies watching Bill receive his blue-and-gold hood was a most rewarding experience for me.

My longest campaign to honor a worthy friend involved Seth Neddermeyer. Seth was the senior author with Carl Anderson of the paper that announced the discovery of the muon. If Anderson hadn't already won (with Seth's help) the 1936 Nobel Prize in physics for his discovery of the positron, they would probably have shared the prize with Harvard's Curry Street for the muon discovery. But the nearly absolute rule of only one to a customer blocked a second Nobel Prize for Anderson and so probably kept Seth from winning what I think he richly deserved.

Furthermore, Seth's proposal at Los Alamos to assemble a supercritical mass of fissionable material by implosion was the key to the successful harnessing of plutonium in a bomb. Without Seth's implosion idea the atomic-bomb program probably wouldn't have ended the war, Japan would probably have been invaded with great loss of life, and after the war scientists would have been accused of wasting enormous resources on an expensive failure. So Seth, I believe, was responsible to a great extent for the high esteem in which science came to be held and for its unlimited financial support in the years following the war.

Carl Anderson and I agreed thirty years ago that Seth should be nominated for membership in the National Academy of Sciences, but in the press of other work we failed to follow through. A few years ago I realized Seth had been overlooked. I proposed him immediately, but the new generation of physicists that by then formed a majority in the physics section of the academy didn't know him. Having struck out with the academy's physics section, I formed a special nominating group. Such groups must have at least twenty-four members, with no concentrations by professional sections or by institutions, and usually have fewer than thirty. I assembled seventy-five. When the final ballots were counted, the academy had elected fifty new members. The fifty-first name on the list was Seth's.

I gave up on the academy then and went after the Department of Energy's Fermi Award, which ranks with the National Medal of Science in prestige. I wrote a strong nomination letter for Seth and was surprised to be asked to serve on the screening panel. I declined because I had a preferred candidate but learned that was the reason I'd been appointed. Since the Fermi Award had almost always been shared before, our panel selected two men: Bennett Lewis, the Canadian who developed the CANDU nuclear reactor, and Seth. At the White House someone decided to save money that year by giving only one award. Without asking about relative merit, they followed our alphabetical order and lopped off Seth's name.

I hoped, in vain, that Seth would automatically become the panel's candidate the next year. The choice wasn't automatic, but Donald Kerst, the inventor of the betatron, fortunately served on that panel and felt as I did about Seth. Seth's name went alone to the White House. President Reagan asked to present the award at the Department of Energy, where no President had ever set foot before, but this time wanted a second candidate. The panel wisely chose Herb Anderson, Fermi's closest colleague. I sat in the second row to watch the President present Seth his award, after which he asked if Seth would like to comment. Seth responded with characteristic modesty: "Someone made a big mistake."

The workings of the award system aren't well known, which is why I've dwelt on them here. I can identify the person responsible for almost every award I've received. If Seth reads these pages, he'll learn that it was I who made the big mistake, of which I'm very proud. I've had a number of quick successes in honors nominations, but they haven't been as interesting as these.

I've said little about religious matters. Physicists feel that the subject of religion is taboo. Almost all consider themselves agnostics. We talk about the big bang that started the present universe and wonder what caused it and what came before. To me the idea of a Supreme Being is attractive, but I'm sure that such a Being isn't the one described in any holy book. Since we learn about people by examining what they have done, I conclude that any Supreme Being must have been a great mathematician. The universe operates with precision according to mathematical laws of enormous complexity. I'm unable to identify its creator with the Jesus to whom my maternal grandparents, missionaries in China, devoted their lives.

Most people start their careers with definite goals: an MBA wants to be a chief executive officer, an Annapolis graduate the chief of naval operations. At each promotion step in the military, and to some extent in business, the rule is up or out. I had the good fortune to enter the academic profession. It has lots of ups—a deanship, a university presidency—but once tenure is attained no outs. I was quickly promoted from the post of instructor to that of full professor, where I have happily stayed. I thoroughly enjoyed my life at the blackboard and taught undergraduate courses exclusively.

Shortly after Jan and I were married, she decided to learn more physics. So each quarter for two years I signed up to teach the course she wanted to audit. The students quickly learned her identity; as a result I got plenty of feedback on the job I was doing. Once, when Jan was standing in line to register, a student suggested she choose someone else's section and take his exams, but attend my lectures. In student evaluations of forty-five physics professors at Berkeley, I was ranked first for several years in the 1970s. Certainly I always felt comfortable lecturing.

The great appeal of the academic world is that no one is looked down upon or forcibly removed if he doesn't pursue promotion. I can't think of anything more rewarding than being a full professor at a great university. Ernest Lawrence, who knew me better than almost anyone, had a strange desire to turn me into an administrator. Fortunately I was able to escape most of his ambushes, sometimes purely by luck.

When he proposed putting me in charge of production at the MIT radar laboratory I had a gallbladder attack. When he and Alfred Loomis wanted me to head the radar countermeasures laboratory, I correctly argued that Fred Terman could do a better job. When Ernest assigned me to direct the proposed heavy-water reactor program, the AEC killed the project. I did direct work on the Materials Testing Accelerator for two years but then managed to escape back into research.

I didn't take Ernest's last attempt seriously. He proposed to the Berkeley regents that I succeed the university president Robert Gordon Sproul. The regents looked me over at a small dinner. Catherine Hearst was friendly, but in sympathizing over the breakup of my first marriage she was telling me that as a Catholic she couldn't support someone who had been divorced. In any event I was sure I couldn't handle the job and wouldn't have accepted it in the unlikely event it had been offered. I've already told how I was offered the job as Ernest Lawrence's successor. The last attempt to lure me into administration came from the leaders of NASA, who wanted me to direct the Goddard Space Flight Center. Through an evening of vigorous lobbying they didn't succeed in convincing me, and I suggested we split the cost of the dinner in compensation.

My family has given me enormous satisfaction and pleasure. After Jan and I were married, we were happy not to be tied down with children. Jan brought luster to the pejorative term "stepmother," and we enjoyed Walt's and Jean's visits. But after six years of traveling and entertaining we decided we would enjoy having children of our own. The evolutionary process that turns individuals into parents found fruition when Donald Luis Alvarez was born on October 15, 1965. We moved into a larger home in anticipation of the arrival of Helen Landis Alvarez on November 9, 1967. Don and Helen have been dream children.

Jan and I have a wonderful marriage because we share so much of each other's life. Jan worked at the Radiation Laboratory and knows everyone I interact with there. I know everyone at Schwem Technology and its contacts outside. Our evenings are rich with talk.

The main difference between my first and present marriage is in communication. During the war I worked under strict secrecy rules for five years; Gerry and I forgot how to share our lives. I was also too frequently away and couldn't afford long-distance calls, which were expensive then. Time has relieved that financial problem, and Jan and I have talked almost every night for three decades, transcontinentally and internationally. Neither of us has ever signed off because the bill was getting high. We talk for as long as we have something to say, and we consider the expense vital. Only from Antarctica, Vietnam, and Egypt did I not call Jan every night.

My father's lifelong experience as a clinician convinced him that longevity is determined by heredity and not much influenced by lifestyle. He lived to the age of ninety-four, his brain working exceptionally well, with only its short-term memory slightly degraded. My mother died at eighty-eight, also in good mental health. Therefore I'm justifiably concerned about my future. One advantage of being a "senior scientist" is that I get many more invitations to serve on boards and commissions than I can possibly accept. I then have the pleasure of picking and choosing. A recent choice is the National Commission on Space, a White House–level commission charged with setting the goals of the U.S. civilian space program for the next fifty years. It met once a month for a year and a half and delivered its report to the President in the spring of 1986.

For several years I have divided my work equally among the study of global extinctions, the inventing of new optical products, and the writing of this book. Most of my creativity now goes into designing a small, lightweight, low-cost stabilized binocular telescope with a wide range of magnification. Pete Schwemin and I are still active in this field after twenty-two years, during which, despite more than a hundred patents, no really successful stabilized binocular has emerged. Pete and I are sure that our new and revolutionary design will lead to the first such success, which Jan and her engineering, production, and sales people will manufacture and sell. All budding inventors are convinced their gizmo will make them millions, but Pete and I are mature inventors, not easily swayed, and our dreams of great financial return from a popular new product should carry weight.

I've also invented a promising new technology that apparently solves the long-felt need to search checked baggage for concealed explosives. All carry-on luggage is searched by X-rays; great efforts have been expended unsuccessfully on methods of searching checked baggage automatically. The FAA is excited about my new technology and will support it financially as soon as some legal problems can be straightened out. And now that this book is finished, I find I'm using the time so freed up to think out all the details of explosives detection.

The rest of my energy is invested in extinctions. The impact theory of mass extinction is gaining wide acceptance. I watch with pleasure as the distinguished Harvard paleontologist Steve Gould runs interference for us. His theory of punctuated equilibrium holds that evolution proceeds by sudden jumps rather than by the slow and steady progress that Darwin had in mind. Our impact theory is just what Steve needed to drive punctuated equilibrium: when species come into equilibrium, only a drastic change in the environment, wiping out most of the well adapted, makes major evolutionary change possible.

My impact theory work convinces me that Milton knew what he was talking about when he wrote, "They also serve who only stand and wait." While many of my friends devoted most of their time to arms control efforts, I looked into extinctions. Soon after my colleagues and I published our impact hypothesis, a group of atmospheric experts at the NASA Ames Laboratory examined it in detail. They confirmed our general conclusions but thought the dust cloud would fall out more quickly than we had predicted. A study that grew out of that work is the now-famous "nuclear winter" paper that proposed that smoke from fires set by exploding nuclear weapons would similarly block out sunlight worldwide with consequences similarly dire. There is as yet no agreement on the correctness of the nuclear-winter scenario, and the latest word is that it is seriously wrong, but it has taken five years of hard work by lots of people to bring our extinction theory to general acceptance.

I was already convinced that there are enough good reasons not to start a nuclear war. But the fact that neither of the two superpowers' nuclear-weapons establishments had thought about the possibility of a nuclear winter has sobered everyone concerned with fighting a nuclear war. What else, they wonder, have they forgotten to think about? The most encouraging feature of the nuclear-winter scenario is that no one has been able to disprove it. It has had a very salutary effect on the thinking of military planners on both sides of the world. There are some indications that it is weakening the Soviet military's long-held belief that nuclear war is survivable. If so, then it may turn out that my rather peculiar way of thinking about new problems will have had an important effect on what I continue to believe is the world's number one problem, the avoidance of nuclear war.

INDEX

Index

Index

Index

Index

Index

Index